CULTURING LIFE

CULTURING LIFE

How Cells Became Technologies

HANNAH LANDECKER

Harvard University Press
Cambridge, Massachusetts
London, England
2007

Library of Congress Cataloging-in-Publication Data
Landecker, Hannah.
Culturing life : how cells became technologies / Hannah Landecker.
p. cm.
Includes bibliographical refrences (p.).
ISBN-13: 978-0-674-02328-4 (alk. paper)
ISBN-10: 0-674-02328-5 (alk. paper)
1. Cell culture.   2. Tissue culture.   3. Biotechnology.   I. Title.

QH585.2.L36 2006
571.6'38—dc22   2006049019

To my parents,

Elizabeth A. Landecker

and

Thomas L. Landecker

CONTENTS

Acknowledgments   *viii*

Introduction: Technologies of Living Substance   *1*

1 Autonomy   *28*

2 Immortality   *68*

3 Mass Reproduction   *107*

4 HeLa   *140*

5 Hybridity   *180*

   Epilogue: Cells Then and Now   *219*

   Notes   *239*

   Index   *272*

ACKNOWLEDGMENTS

It is with great pleasure that I set out to recognize those who have aided, encouraged, shaped, and financed work on this book. Joseph Dumit, Michael M. J. Fischer, Evelynn Hammonds, and Evelyn Fox Keller read and engaged the first version of this work, and to them I extend my warmest thanks. Michael Fischer has been an extraordinary mentor and friend. I would like to thank Leo Marx for years of discussions on literature and writing and many walks across the Longfellow Bridge. Hugh W. Brock taught me to love biology—both the living things and the science—and his teaching has stayed with me well beyond my year in his laboratory.

I had the good fortune to be part of rich intellectual communities in Boston and Berlin, at MIT and the Max Planck Institute for the History of Science. Colleagues in these places, and in the networks of history and anthropology of science, have in conversations, comments, and their own work contributed to the final shape of this book. Sarah Franklin, Sarah Jansen, Janina Wellmann, Jean-Paul Gaudillière, Hans-Jörg Rheinberger, Nick Hopwood, Michelle Murphy, Matt Price, Nick King, Aslihan Sanal, Stefan Helmreich, Stephen Pemberton, Adriana Petryna, Debbie Weinstein, Susann Wilkinson, Michael Kyba, Rebecca Herzig, Dan Harris, Barbara Harris, Fabrizia Giuliani, and Roberta Bivins have all contributed, one welcome way or another, to my writing and thinking. More recently, I have enjoyed the welcome of the faculty and graduate students of the Department of Anthropology at Rice University; thanks particularly to Jim Faubion and Michael Powell for their comments on the manuscript. George Marcus, Kirsten Ostherr, Scott McGill, Sarah Ellenzweig, and Jennifer Hamilton have made Rice a wonderful place to work over the last few years. Rich Doyle is a constant creative presence from afar. Tracy St. Claire, Cynthia Bohan, Sandra Chalmers, and Angela Ashton have been there from the beginning.

A fellowship from the Center for Medical Humanities at the University of Texas Medical Branch at Galveston allowed me invaluable peaceful time working in the UTMB library among the sleeping medical students. The hospitality of the History of Science Department at Harvard and the Program in Science, Technology and Society at MIT in hosting me as a visiting scholar in 2005 was essential to finishing the book. A fellowship from the National Endowment for the Humanities supported research and writing for Chapter 2 (but any views, findings, conclusions, or recommendations expressed do not necessarily reflect those of the NEH). In addition, I would like to thank Mazyar Lotfalian and Melissa Cefkin, in whose apartment Chapter 3 was drafted.

Portions of Chapter 1 are revised from their original publication in "New Times for Biology: Nerve Cultures and the Advent of Cellular Life in Vitro," *Studies in the History and Philosophy of Biological and Biomedical Sciences,* 33:667–694. Chapter 2 includes revised passages first published in "Building 'A new type of body in which to grow a cell': Tissue Culture at the Rockefeller Institute, 1910–1914," in *Creating a Tradition of Biomedical Research: Contributions to the History of Rockefeller University,* edited by Darwin Stapleton (New York: Rockefeller University Press, 2004), 151–174. Chapter 4 contains revised passages from "Immortality, *In Vitro:* A History of the HeLa Cell Line," in *Biotechnology and Culture: Bodies, Anxieties, Ethics,* edited by Paul Brodwin (Bloomington: Indiana University Press, 2000), 53–72.

My grandfather Patrick DeBurgh would have been a part of Chapter 3, but for an accident of history. I would like to thank him for sharing his memories of midcentury bacteriology. Tom and Elizabeth Landecker, Jim Landecker, Kath Becker, and Ted and Anne Kelty have been constantly supportive through years of reports on what must have seemed like an endless project.

Chris Kelty read and discussed every word and idea, for years on end, and remained interested. He gets more than thanks; he gets all the love there is. And Ida, who was born as this book was near completion, provided the best deadline anyone could ever work toward.

There is in science, and perhaps even more so in history, some sanction for the belief that a living thing might be taken in hand and so molded and modified that at best it would retain scarcely anything of its inherent form and disposition; that the thread of life might be preserved unimpaired while shape and mental superstructure were so extensively recast as even to justify our regarding the result as a new variety of being.

—H. G. WELLS

A tissue is evidently an enduring thing. Its functional and structural conditions become modified from moment to moment. Time is really the fourth dimension of living organisms. It enters as a part into the constitution of a tissue. Cell colonies, or organs, are events which progressively unfold themselves. They must be studied like history.

—ALEXIS CARREL

INTRODUCTION:
TECHNOLOGIES OF
LIVING SUBSTANCE

This book presents a history of tissue culture, the practice of growing living cells outside the body in the laboratory. At the same time, it tells the larger story of twentieth-century ideas and practices of plasticity and temporality of living things: Living things may be radically altered in the way they live in space and time and thus may be harnessed to human intention. This history highlights our human relationship to living matter as one structured by the concept of life as technology. Examining five central developments in the use of cultured cells over the twentieth century, I illustrate how novel biotechnical objects such as endlessly proliferating cell lines affect concepts of individuality, immortality, and hybridity.

In 1890, the biologist Jacques Loeb wrote to physicist Ernst Mach that "the idea is now hovering before me that man himself can act as creator, even in living nature, forming it eventually according to his will. Man can at least succeed in a technology of living substance."[1] Although there are many bio-words, from the original biology to biomedicine to bioscience to biotechnology, I have written a history of tissue culture that is particularly relevant to understanding "technologies of living substance," a phrase more inclusive and thus more useful than the contemporary term "bio-technology." The word's present connotations point to biotech-nology as an economic and scientific phenomenon that began in the 1970s with the power of recombinant DNA and its industrial applications[2] and the restructuring of the economic and legal con-ditions of the life sciences, which brought an influx of private cap-ital into academic biology.[3] I invoke Loeb's phrase here to indicate the realm of biological technologies that existed well before the 1970s but outside the etymology of the word "biotechnology" it-self, which traces back to earlier visions of organisms such as bacte-ria and pigs as factories for the production of valuable substances.[4] This is an effort not to redefine or define biotechnology but to pose the broader question of how humans have come to regard and in-teract with living matter through the framework of life as tech-nology.

This assumption of living matter as technological matter is con-stitutive of life today, in terms of both how it is lived and how it is concretely approached, handled, and manipulated. Living cultured cells are today used widely in research programs of all kinds and also serve as productive sources of biological molecules for phar-maceutical research and therapeutics, the food industry, and bio-medical research. Cells may be cultured short term as proxy diag-nostic bodies for the patients from which they have been extracted; or they may be cultured long term, as so-called cell lines or perma-nent *in vitro* populations of self-replicating somatic cells. Cells from all manner of organisms, from plants to insects to animals to hu-mans, constitute a substantial biomass present in the laboratories of the world, a living material base for contemporary life science.[5] The science, technology, and economic productivity of living mat-ter depends on the productive and reproductive capacity of cells to continually make more of themselves while also generating large volumes of the biological materials of research—enzymes, antibodies, DNA, RNA, viruses. Thus the contemporary cell is also an important economic entity, patentable and productive.

Anthropologists of science and technology have noted the living cell as scientific and technical object in ethnographies of Western biomedical and biotechnical settings. From amniocentesis to the establishment of cell lines, the substance of the human body is now routinely maintained alive outside the body.[6] These disembodied, productive, replicating cells that are derived from human bodies but live in laboratories are new but also newly normal. It takes an anthropologist in the laboratory to note the strangeness of what has become quickly routinized or banal to its practitioners. Cultured cells are characteristic of the present of the human condition, they function within well-established systems of labor and exchange, they are normalized in and by these systems; yet they also represent profound and recent change to a new state of being, as routine tools, alienable commodities, and sites of production. One does not have to be an anthropologist to see that a shift has occurred, new forms and practices of life have appeared, and humans regard human biological matter differently than they did before.

The life form of the cultured cell is a manifestly technological one: It is bounded by the vessels of laboratory science, fed by the substances in the medium in which it is bathed, and manipulated internally and externally in countless ways from its genetic constitution to its morphological shape. Its existence bears little resemblance to the body plan or the life span of the organism from which its ancestors were derived. Contemporary life in this particular form is something that exists and persists in the laboratory, the niche of science and technology.

Despite its relative novelty in historical terms, this state of life has quickly become normal, imbuing scientific objects such as cell lines with an aura of inevitability or, ironically enough, with an air of natural existence. How is it that life, once seated firmly in the interior of the bodies of animals and plants, came to be located in the laboratory? At what point did living matter get extracted from and stripped of the individual forms of organisms? Further, why did the cells of humans become incorporated into the research biomass along with those of other organisms, and how do the lives of such human-derived objects affect the concept of the human subject? How did life, including human life, take this contemporary disembodied, distributed, continuous form? The question of where tissue culture came from is not only one of origins but also of conditions—of what makes it possible for these biotechnical things to exist in these detached, transformed ways.

A glance at newspapers or scientific journals today indicates that both scientists and nonscientists are now thinking a great deal about cells and these questions of their technical form, in contrast to ten or twenty years ago when they were only interested in genes. Thus, a history of cells in biology is timely.[7] Indeed, the contemporary life sciences are now undergoing a shift in emphasis away from such exclusive privileging of the DNA base sequence of coding regions of the genome.[8] Whether we identify this shift as postgenomics, metabolomics, proteomics, epigenomics, or stem cell biology, the linearity of the central dogma—DNA makes RNA makes protein—is being corrected by the elaboration of other complex temporal and spatial relationships between biological molecules that are ordered by the structure and function of the living cell.

The cell is making a particular kind of reappearance as a central actor in today's biomedical, biological, and biotechnical settings. From tissue engineering to reproductive science, culturing the living cell outside the body has become increasingly important to

4

making new biotechnical objects. At the beginning of the twenty-first century, the cell has emerged as a central unit of biological thought and practice.[9] The cell has deposed the gene as the candidate for the role of "life itself." The realization of the cell's role in coordinating the molecular processes so painstakingly detailed by biochemistry and molecular biology over the twentieth century shifts our focus to the cell as dynamic entity: "Chemistry made into biology."[10]

At the same time, this emergence appears to be something of a reemergence. During the time I was writing this book, I was on my way to class to lecture about Jacques Loeb's 1899 discovery of artificial parthenogenesis when I saw a newspaper headline announcing the successful use of this very technique by a biotechnology company to spur monkey eggs into embryonic development without the presence of sperm. Loeb had altered the concentration of salts in the water surrounding sea urchin eggs, thereby pushing them into cell division and early embryonic development without the presence or fertilizing action of sperm. He coined the term "artificial parthenogenesis" to describe the process. Advanced Cell Technologies was using basically the same approach. Loeb interpreted the ability to begin development by altering salt concentrations as evidence of the physicochemical basis of life; for the company, it represented an opportunity to make embryonic stem cells without having to go through the fertilized egg or the embryo made by nuclear transfer. A parthenogenic embryo cannot grow to a full adult under any circumstances, so this technique provides a kind of ethical bypass around the issue of destroying human blastocysts to disaggregate them into stem cells.

What should we make of the reappearance of techniques such as artificial parthenogenesis a century later? Historians of science familiar with late nineteenth-century and early twentieth-century biology will see in this concern with the cell not just a return of the conceptual emphasis given the cell as the "elementary organism" and the most basic form of life but also the reappearance of techniques of cellular manipulation such as artificial parthenogenesis and of terminology such as "totipotency."[11] After decades of attention to other biological entities, particularly The Gene, such a return to the livelier, complex, and indeed entirely more personable entity of the cell seems refreshing. One senses a certain amount of glee in observations that a refigured cell—"one of biology's oldest and most classical points of reference to describe life's commonalities"—has "burst back on the scene."[12]

Such language of disappearance and return might imply that the cell went somewhere while everyone was working on genes and talking about codes and programs—into hiding, into the back of the freezer, into obscurity—and it has now returned. The philosophy of Georges Canguilhem offers a more complex version of recurrence: The present of a science is always going through critical self-correction, and those elements of the past that seem constitutive of the present condition of a science change accordingly.[13] Paying attention to the gene in the context of the cell, and to the cell as more than a passive backdrop to the machinery of DNA, RNA, and proteins, is characteristic of today's self-correction; and such attention brings cells to center stage. This correction is driven as much by inquiry into the true nature of the cell as by the realization that the manipulation of genes is ineffectual without the manipulation of cells.

If we look at the historical literature, it would seem that the cell was of little importance after the rise of genetics and molecular biology in the 1920s and 1930s; extensive or systematic historical inquiry into the biology of cells for the most part ends at the point at which cell theory was proposed, argued over, and largely accepted in the nineteenth century.[14] Surveys of the history of science or of the life sciences in the twentieth century have chapters on genetics and evolutionary theory.[15] In François Jacob's history of biology, for example, a chapter entitled "The Cell" covers a period of time at the end of the nineteenth century; and chapters entitled "The Gene" and "The Molecule" cover later work."[16] Such was the perspective on the previous centuries from the vantage point of biological science in the 1970s. But the cell has always been there in the scene even when the gene or the molecule seemed the central player. Present conditions make the cell's role more visible and thus make it possible to track and tell elements of histories that had previously faded into the background of other dominant narratives or were visible only as marginal counternarratives.

Because of the recent prominence of stem cells and cloning in public discussions of the life sciences, many readers may expect this book to be a history of these particular technologies. Although the history of tissue culture is central to understanding these modes of manipulating cellular life outside the body, it is not the same as a history of developmental biology and reproductive science. My aim here is simultaneously more specific and more comprehensive. The narrative sticks to tissue culture, while arguing that the history of cell cultivation is the history of an approach to living matter that encompasses any specific example of biotechnical innovation such as stem cells. Understanding this approach means learning about the context and conditions that make new cell technologies such as stem cells possible.

This book articulates one of the several pasts of contemporary biology and biotechnology, as its present undergoes change and the unit of the cell becomes more scientifically, technically, philosophically, and economically important to how living things are thought about and manipulated. It does not chart a reemergence for the cell or for ideas and practices of its plasticity; instead, it illustrates the constant presence, development, and results of the questions and explorations begun a century before. They seem to be returning, but actually they never went away, even as the center of the life sciences was occupied with other things; in fact, the often invisible infrastructural conditions for today's developments were being set even as genetics and molecular biology were ascendant. It is this constant presence of the cultured cell over the twentieth century, and the development of techniques for growing, keeping, and reshaping these cells and their lives, that establishes the conditions for the current traffic in cells that undergirds research and commerce.

Thus it is not coincidence but recurrence that lies behind the century-long cycle of appearances of artificial parthenogenesis in the newspapers. Today's apparent return of the cell is continuous with the questions and explorations of the cell and its plasticity at the beginning of the twentieth century. A genealogy of plasticity structures today's experimental probing of the manifold potentiality of living matter and the practical experimental milieus in which cells are made to live. Recurrence is not a reappearance of the same or a return of the repressed, but a set of emphases with which to recognize a genealogy that has always been there. In this account, the details of the development of a particular technique—tissue culture—are ordered by emphasis on those practices that exploit and explore the plasticity of living things. The remainder of this chapter introduces the theme of plasticity and temporality, explains how each of the chapters elaborates this central theme, and discusses the methodological challenges of writing histories of twentieth-century life science.

Plasticity and Temporality

The plasticity of living things—and a biological science concerned with life's plasticity—is the historical starting point and the frame for this telling of the history of tissue culture. In 1895, H. G. Wells, a young biology instructor just launching a career as a science fiction novelist, wrote a short newspaper piece called "The Limits of Individual Plasticity" in which he criticized the fatalism of hereditary thinking. He pointed to new developments in transplant surgery, blood transfusion, and hypnosis that seemed to indicate that living matter was highly malleable and that human intervention could mold and modify it radically. In words that he placed in the mouth of Dr. Moreau shortly thereafter in the novel *The Island of Dr. Moreau*,[17] he asked whether a living thing could be "taken in hand" and so changed that it would constitute a "new variety of being." The continuity that binds the original living thing to the one recast by human intervention is an unbroken "thread of life." This thread of life might be preserved unimpaired even as the substance of the living thing, the matter of its body, is totally reshaped: manipulated form, same life.

Wells does not so much provide a definition of plasticity as pose a question as to its limits. He was clearly ambivalent about the potential answer to how extensively living things could be reshaped: *The Island of Dr. Moreau* ends with the constructed human-animal monstrosities quickly reverting in form and behavior to their inherent animal natures, despite Dr. Moreau's efforts to make them human. The humans mirrored in turn by these animals are no more able to transcend their bestial evolutionary origins. The novel experimented in fiction with living substance much as Wells saw it being probed by the actual biological science of his time: How far could one go in altering the substance of living matter before the thread of life was broken? He was asking what the difference was between a partly man-made organism and a naturally occurring one. Was the nature of life set from birth, inherent to matter, or could it be changed by human intention?

The biologist Jacques Loeb, Wells's contemporary, was less ambivalent in voicing a similar, nonfictional ambition for human intervention. Loeb saw biology not as a science that should stand back
and observe nature but as one that should take living substance in
hand and reshape it to particular ends, an approach historian Philip
Pauly has called Loeb's "engineering ideal" for life science.[18] The
question for Loeb was simultaneously one of the limits to organisms' malleability and of the boundaries to human action in nature—the question of the very possibility of reforming or re-creat-
ing living matter as technology.

Although neither Wells nor Loeb are directly involved in the history recounted in this book, their questions are presented here as
an entry point into the concept of plasticity. Plasticity is an idea of
living matter that is also a practical approach to it: Substance may
be radically altered without causing death. Plasticity is the ability
of living things to go on living, synthesizing proteins, moving, reproducing, and so on despite catastrophic interference in their constitution, environment, or form. Wells's novel reminds us of an important second quality of plasticity: Although living things can be
radically manipulated, part of the particularity of biological plasticity is that biological matter may change or react to intervention in
totally unexpected ways.[19] From Jacques Loeb's ambitions for biological science, the third quality of plasticity can be teased out:
not just the inherent ability of living matter to adapt flexibly and to
live through shock and rearrangement, but also its capacity to *be*
changed by humans. Biological plasticity thus has three facets. It is
the plastic quality of living matter; it is the fundamental unpredict-
ability of life even in the face of human intention and construction;
and it is a form of practice that is part of modern biological science. All three of these elements of plasticity are important to understanding the course and significance of tissue culture in the
twentieth century.

The first establishment of tissue culture in 1907 was very much part of the questioning of the limits to the malleability of living substance that Wells and Loeb were participating in, and the concomitant excitement over the expansion of the limits of experimental intervention. The ability of cells to grow without the body that they constituted was the first shock to assumed limits of cellular autonomy and bodily integrity; from then on, the entire history of the practice in the twentieth century could be generally described as a series of realizations of cells' abilities to withstand and live through a variety of rude manipulations, from extracting them from their bodily context to fusing them together artificially. These "realizations" often took the physical form of a cell culture living in a particular and unexpected way after intervention; its manifest existence altered the meaning and possibility of fundamental categories of biological and cultural thought, from autonomy to immortality to hybridity.

Also characteristic of the development of tissue culture is the intertwining of practices and concepts of biological plasticity with those of biological temporality. The reshaping of cellular living matter has been linked step by step to a manipulation of how cells live in time. For example, immortality as a technical term in biological science made its appearance in the first decades of the twentieth century with the claim that cells freed from the bounds of the body are also freed from the limits of the originating organism's life span. Techniques of plasticity are modes of operationalizing biological time, making things endure according to human intention.

In the history of tissue culture, plasticity, the pushing and pulling of biological things to live in different shapes and spaces and times, is achieved in many cases by a manipulation not just of the cell itself but also of the medium in which it lives. Attention to the cellular medium is a central part of this account of how cells came to live in laboratories. This manipulation of cells and medium—living

substance and its conditions of life—was often directed at controlling biological events of development, infection, and reproduction
such that they happened when the scientist needed them to happen. In practical terms this entailed attention to the fluids, gases,
nutritional substances, and surrounding apparatus that kept cells
alive in the laboratory. Being able to manipulate substance in each
of these cases was closely tied to learning how to manipulate cells
as well as cells via their medium. One tissue culture practitioner
commented in 1916, "Through the discovery of tissue culture we
have, so to speak, created a new type of body in which to grow a
cell."[20] "Medium" thus refers to the liquids immediately surrounding a fragment of cultivated tissue and in an expanded sense to the
whole apparatus supplied by the laboratory to replace the body.

The manipulation of medium was, and is, in turn tied to the manipulation of cellular time. A history of such banal things as nutrient media is a necessary part of any history of the manipulation of
biological time. Experimenters struggled with questions such as,
What conditions are best for keeping cells alive permanently? What
conditions allow clonal lineages of somatic cells to be established?
The poet and immunologist Miroslav Holub once observed that organisms are improperly organized for scientific technologies.[21] Although he was speaking mostly of the problem of size, this is also
the problem of time, because the time of the scientist and that of
the experiment do not necessarily square with that of the organism
in question. Temporal reorganization of the organism as tissue cultures took the form of techniques directed toward the event, making processes happen faster or slower in the desired fashion, or establishing continuity through clonal lineages that otherwise would
not have existed. Any genealogy of plasticity, then, is also one of
temporality.

The manipulation of biological form and time is important to the approach to living matter found in biomedicine and biotechnology today. From metabolic cycles to life spans to growth rates, reshaping form is also a reshaping of how life exists in time, as if contemporary scientists have taken Ludwig von Bertalanffy's insistence that structures are "slow processes of long duration" and functions are "quick processes of short duration" as advice on how to tinker with life.[22] Cultured cells are now commonplace, and their current form is one that is characteristic of assumptions of the plasticity of living matter fundamental to today's biosciences. The material, manifest change in how living things exist in the laboratory is a shift in the understanding of biological form and time, and simultaneously a shift in concepts of what technology can or might be, now that it encompasses living matter. These shifts have obvious philosophical and cultural implications,[23] and these may become more fathomable by understanding the history of today's plastic, temporally malleable life.

The concepts and practices described in this book are part of the "revised spatialization and territorialization of living processes" that produce new knowledge of bodies and organisms while transforming the very objects of scientific knowledge and technical production themselves.[24] Put simply, the cell and the body of science and culture are not the same after tissue culture. In a review of a book about investigating hormones in tissue culture, scientist Robert Pollack observed that his dictionary "defines a hormone as a substance found in some organ of the body and carried by a body fluid to another organ or tissue, where it has a specific regulatory effect." But cell culture "frees the cells of the body from such distinctions as internal and external, tissue and organ, and makes them all equally accessible to experimental manipulation of their capacities to grow and to differentiate." How then, he asks, "can a hormone be defined without a body?"[25] This simple question encompasses a world of change in biological and biomedical thinking. It is not just the cell that is transformed by life's technical reformation but also the body and its processes. This history of tissue culture is not designed to be comprehensive. It contributes to an understanding of the transformation of the cell as a technical object—a transformation that in turn has altered the science that takes the cell as its object and subject of research.

Five Chapters in Twentieth-Century Cellular Life

The disembodiment and redistribution of living matter from bodies to laboratories is the central story of this book, which traces the *in vitro* life of cells over the twentieth century from approximately 1907 to 1970. The book is divided into five chapters, each centered on an episode in which cell cultures were, as Alexis Carrel phrased it, "events which progressively unfold themselves." In the course of each of these events, shifts in the manipulation of living cells and their surrounding medium resulted in a manifestly different way for cells to live in space and time. With each of these rearrangements of living matter, concepts and practices of individuality, bodily integrity, life spans, the separation of species, and the human body as biomedical research object were fundamentally altered. The book ends with an epilogue that takes the form of a question: How can the history presented here alter the analysis it is possible to do of contemporary developments in the biosciences?

The shift from nineteenth-century *in vivo* experimentation in the body of whole animals to *in vitro* experiments on pieces of the body kept in artificial conditions in the twentieth century is the beginning and context of this story. Chapter 1, "Autonomy," opens in 1907 with the work of Ross Harrison, an American embryologist who first demonstrated that tissue fragments from whole bodies could live *in vitro* for weeks at a time, as long as they were immersed in a suitable medium and protected aseptically by their glass enclosure. Scientists experimenting with the new method encountered novel and uncanny sights such as an isolated heart muscle cell pulsating by itself. This unexpected level of autonomy of cells in relation to the body was greeted by observers with both disbelief (can it possibly be true?) and excitement (if it is so, imagine the experiments that can be done!).

Harrison established the beginnings of a technique that would come to be called tissue culture, a technique whose temporal design simultaneously made it an event. In Harrison's work, this was the event of embryonic nerve development, brought out of the dark inside and obfuscating context of the solid complex body and into the simplified transparent technical body where it could be continuously observed. For some life scientists, this experiment was significant because it said something definitive about nerve development; for others the finding had the more nebulous but profound quality of proving the *possibility* of observing internal bodily events without the body itself—observations that had previously been assumed to be impossible, if considered at all. Harrison's work was picked up by others not because they wanted to observe embryonic development in the same way but because they wanted to explore the possibilities of life *in vitro* and repeat the event of life over time. This was the case with Alexis Carrel, a Franco-American surgeon who quickly switched the focus from the specific findings of Harrison's careful experiments to the method and its possibilities.

Whereas Chapter 1 is concerned with the new autonomy for cells extracted from the confines and shape of the animal body, Chapter 2, "Immortality," turns from disembodiment to continuity, examining the work of Alexis Carrel in establishing what he called the "permanent life" or potential immortality of cells in culture. Unlike Harrison, who did not observe or seek to observe the excised cells dividing, Carrel was deeply concerned not just with the temporary sustenance of life for a matter of weeks but also with having the cells reproduce and continue to reproduce in their new *in vitro* form. He focused on controlling how they lived in time through manipulating the medium in which they were kept. Gaining control over the cells and their medium was for Carrel a means of manipulating physiological duration and producing an immortal object.

Carrel chose as his means of demonstration of immortality a culture of embryonic chicken heart cells, which came to be known as "the immortal chicken heart" and had what one of Carrel's assistants called "the most remarkable career ever enjoyed by chick or part of chick" from 1912 to 1946.[26] Here the chapter turns to a subplot of this book: how these technical manipulations of life relate to the reconfiguration of concepts of very broad cultural salience, such as immortality. Paying close attention to the practical culture of Harrison's and Carrel's laboratory materials and techniques allows the examination of laboratory practice as a simultaneously "cultural practice" of life science in the early twentieth century.[27] In these accounts, one sees the fathoming of a new kind of object or form of life—an immortal tissue or cell—and, simultaneously, a narrative of life that never loses sight of the technical mediation which constitutes the very grounds of that life form's possibility.

This glass and apparatus–bound immortality is a window onto what Philip Pauly has called "biological modernism"; another way to see the shift from *in vivo* to *in vitro* experimentation is this increasing emphasis on artifice in biological science.[28] Scientists and their publics alike had to come to terms both with new forms of life and with new modes of science—in this case, a biology focused on shaping living matter in space and time to exist and persist in a specific laboratory form that would never be found in nature. The very possibility of such technologically mediated life forms was in itself seen as a result or finding of scientific work.

The new modes of disembodiment and continuity for cultured cells entailed changes in the way the cells of a single body could be distributed. The implications of this change in possibility for the distribution of cells became pronounced only when human cells began to be cultured on a large scale in the 1940s. Chapter 3, "Mass Reproduction," concerns the development of the human cell as a means of large-scale production of viruses in the effort to find a vaccine for polio in the 1940s. This was made possible by, and helped to further the transformation of, tissue culture from a localized, idiosyncratic practice pursued on a small scale in a handful of laboratories to a much more standardized, widespread practice. In the course of this transformation, human cells were cultured routinely for the first time, and the disembodied living human cell as experimental subject and productive technology came into being.

Although the story of the development of the polio vaccine in the mid-twentieth century is well known, in this chapter it serves as the background to the development of human tissue culture as a means to grow viruses and the resulting fundamental reorganization of the human body as research subject and research object. The work of virologist John Enders forms the core of Chapter 3. After Enders's experiments with tissue culture systems, cultured cells could be used as a means of producing large quantities of virus particles, as a mode of diagnosis in which a patient sample could be tested to see whether it infected a cell culture, or as a system for studying the effect of virus infection on cells.

The work of Enders is deeply entwined with the story of the first widely used human cell line, HeLa, established in Baltimore in

1951 by George and Margaret Gey. Chapter 4, "HeLa," focuses on the disembodied, distributed continuous form of life that was and is the HeLa cell line. Developments in the ability to clone individual somatic cells in culture and freeze cultures in suspended animation for long periods of time in the 1940s and 1950s once again demonstrated the amazing resilience of cellular life in the face of drastic manipulation. Cells could be isolated and then coaxed to divide. Up to that point cultures were always a heterogeneous mix of cells from the originating tissue explant, and half a century of debate since Harrison's original experiments had failed to decide the limits to the cellular autonomy that he had begun to explore. Once a population was descended from a single cell by mitotic division, it came into a genetic individuality that somatic cells had never possessed before. At the same time, development of techniques for freezing cell cultures such that they would continue to live and reproduce upon thawing meant that cells could be maintained in stable, long-term suspended animation and shipped long distances with ease. With freezing and cloning, scientists in different times and places could say, with new meaning, that they were all working on "the same" cell.

In many cases, this "same" cell was a HeLa cell, given its early ubiquity in tissue culture practice. Because this cell line was derived from a person and bore the traces of that person's name and identity, the cell line continued—and continues to this day—to be referred to in relation to the person Henrietta Lacks from whom the original biopsy tissue was taken in 1951. In a reprise of some of the themes of Chapter 2, scientific, science popularization, and media accounts of HeLa and Henrietta Lacks are read in this chapter as one continuous discourse in which scientists and their publics alike attempted to fathom the new conditions of possibility for humans and human bodies in biomedical science. The singularity of one named person's body becoming distributed in this way, around the world and across the decades, outliving the unfortunate tissue donor and even the scientists who established the cell line, was the century's second reconfiguration of immortality in relation to cultured cells.

Disembodied, distributed continuity loosened cells from bodies, human tissue from persons, and biological things from the given time of life spans and other biological cycles. However, this alienation from body and life span was by no means the bounds of plasticity of living cells. Chapter 5, "Hybridity," focuses on the subsequent recombination of cultured cells in experiments of cell fusion in the 1960s. It was during the course of these experiments, which fused cultured somatic cells from different animals and often different species together, that biologists first realized that the boundaries of species integrity signaled by infertility, and the boundaries of organismal individuality signaled by immune reactions and rejections of transplanted organs, did not apply to the deep insides of organisms—at the level of the interior of their cells. In cell fusion, it was not only the cytoplasms of two cells that were fused but often also the nuclei, leading in many cases to a fully functional hybrid cell that could itself reproduce in culture, sometimes indefinitely.

Many concepts and practices of biological sameness and difference are contained in the hybrid, and these were fundamentally altered with cell fusion. Previously constrained to a handful of organisms produced at the edge of species lines, such as the mule, or the production of plants by grafting or manipulating pollination, the hybrid after cell fusion was a much more radical and surprising juxtaposition of biological difference. This was a realization of the internal compatibility of organisms that came before large-scale gene sequencing, sequence comparison, and the realization of homology across species at the level of nucleotides. This period also saw the emergence of a language of the "reconstituted cell." Various fragments—cytoplasm alone, a single chromosome bounded by a membrane, an isolated nucleus—were recombined with whole cells or other "viable" fragments to put a functioning cell back together again. There was no question of such entities ever occurring in nature; these were entirely artificial constructs that opened up the inside of the cell as a space of juxtaposition and experimentation.

Because they are endlessly proliferative, there is no proper end point for a history of cultured cells. Stopping in the 1970s, "Hybridity" as the fifth and final chapter is less of an end point than a certain realization of the artifice, plasticity, and technology that Wells and Loeb envisioned as the future of the human relationship to living matter as well as of the "catastrophic" situation that Georges Canguilhem (following Kurt Goldstein) saw in life subjected to the milieu of the laboratory.[29] This life of cells was one of being taken apart and reconstituted. The integrity of the cell in itself as a separate entity bounded by a membrane and kept apart from other cellular bodies by the bounds of individuality or species immune mechanisms was bypassed. The thread of life continued, and a new hybrid arose from the merger of viable parts: a catastrophic, artificial, and new variety of being, a technology of living substance.

Finally, a short epilogue demonstrates how the genealogy of plasticity and temporality can be used for studying life science and biotechnology in the present. This can be seen as a version of *Wide Sargasso Sea* or *Grendel*—retelling some of the repetitive stories of contemporary biotechnologies not exactly from the point of view of the ignored and banal elements of the freezer or cell media, but via their infrastructural role in the making of biological things. The late twentieth-century tale of cloning is retold not as an event that foreshadows the ability to clone humans or even clone human organs or transgenic sheep that produce human blood clotting factor in their milk. Rather, it is a tale of cell science and how its attendant manipulations produce what H. G. Wells might have called a "new variety of being." Here the work of the first five chapters in outlining the conditions of life for cells in the twentieth century refigures the kind of analysis that is possible to do with particular instances of cellular manipulation in biotechnology today.

Materials and Methods

The narrative I present picks a thread through the twentieth century without writing a biography of a particular scientist, analyzing a particular written work, theory or controversy, or profiling the work of a particular laboratory or institution; it cuts across these domains with an infrastructural approach.[30] I identify genres of technique—those concerned with plasticity and temporality—common to biotechnological objects of apparently disparate kinds. How does one go about detailing the constitution of cells as technologies, the material process of their separation and alienation? The short answer is to consult the materials and methods sections of decades of research papers.

Although I consulted many archives and engaged many practicing scientists in conversation while researching this book, the route I followed through the mass of twentieth-century life science was derived for the most part from the published record. This choice requires some explanation to counter the assumption often found in ethnographic and historical scholarship that interviews and archives are somehow closer to reality than published papers. Accompanying this assumption is the perception that it is more authentic—or at least more recognizable as research—to have accessed the experience of the person or documents not processed for public consumption. Published papers are, as they say, secondary sources.

However, the scale of twentieth-century and contemporary scientific publishing presents a challenge to anthropologists and historians that escapes traditional ethnographic and archival methodologies. Before World War II, and the spread and standardization of tissue culture practices, only a handful of laboratories were focused on developing and exploring the technique itself; but even in this period the literature of tissue culture was huge. When Margaret Murray, one of the founding members of the Tissue Culture Commission in 1946, took on the task of preparing a bibliography of tissue culture, she predicted that 2,000 or 3,000 references would be contained in the projected compilation. In the end, the bibliography published in 1953 contained 23,000 titles published between 1884 and 1950 cross-indexed into 100,000 entries.[31] After 1950, tissue culture became much more widely practiced and the volume of literature rose accordingly; now a single comprehensive bibliography of the subject would be unthinkable. Facing such numbers, the brute effort of reading thousands of short scientific papers, technical manuals, and conference proceedings seemed the only way to enter into the production of these objects and concepts as they occurred, to get at the texture and density of a whole field of activity.

Rather than simply calling it brute effort, however, this approach should be explicitly marked as one that takes scientific literature as a primary informant or source. Citation indexing has taught us that any scientific body of literature is a collective representation of the scientific community, and any mass of journal articles is an entity with its own dynamics.[32] I was continually frustrated by trying to comprehend the movement and dynamics of this entity through single interlocutors, living or dead, mostly because I was trying to understand the movement of a field, a mass of research, rather than the actions of any one person or laboratory or institution. In the anthropology and history of biotechnology, there are many case studies and few synthetic works that give us ways to weave together the case studies into some more comprehensive account of what has happened in the twentieth century. I did not want to do a case study and then generalize; I wanted to do *highly specific empirical work on the general*. This might sound like a contradiction in terms, but it is what led me to this reversal in which I felt that the literature was my primary source and the interviews and archival material were secondary instances of the individual processing of that literature.

The focus in my research was on the places where people recorded what they did and how they did it. I looked for narratives of material action, taking the whole mass of literature as "the work" to be analyzed. To some readers, my concentration on the depiction of action on matter may make it sound as if these experiments were performing themselves. Where are the people? Who did this work? Narratives of personalities and discoveries now abound for twentieth-century science, from *The Double Helix* onward, but this is a different kind of book.[33] Throughout, the focus is on the human approach to living matter contained in these accounts of experiment.

An empirical focus on statements rather than individuals or events; the claim of a genealogy of plasticity for contemporary life sciences; the identification of a shift to life apart, outside rather than inside the body—all of these points invoke Michel Foucault's analysis of "life" in *The Order of Things*.[34] However, a discourse analysis of "life" after 1900 such as Foucault produced for 1800 is impossible if pursued on exactly the same terms. Already in the nineteenth century scientific discourse was of immense density, and individual thinkers and landmark works have only become less adequate means of fathoming scientific activity and its changes. This is why an explicit methodological approach to the large scale of twentieth-century science is needed. Although developments as momentous as the nineteenth century's rise of evolutionary theory have occurred in how we think of living things and act upon them in the twentieth century, "there is no molecular Darwin" or any other dominant figure to provide any unifying logos for us.[35] There is no *Origin of Species* to read for the transformations of narrative in and by theories of life.[36] In fact, it is quite possible that momentous change can be traced not through any single work but through thousands of decidedly little known publications with titles such as "Molecular Growth Requirements of Single Mammalian Cells." There is no disputing that life in its material, conceptual, and narrative existence has been transformed by the vast enterprise that is Western biomedicine and biotechnology. Both life as humans live it in terms of bodies and health and the concepts and the objects that are at the center of our sciences of living things have been radically altered.[37] How to fathom this transformation through historical research that respects the scale and multiplicity of modern scientific work is the challenge.

Biological science and the history of biological science have both changed since Foucault and Canguilhem; this too is part of recurrence. The change most pertinent to the methodology and subject matter of this book is a focus on practice—not just what scientists think or write but also what they do and the materials they work with. Tissue culture, as its earliest practitioners observed, almost immediately came to denote both the material thing and the field of knowledge produced by work with that thing; its history is both a history of ideas and the material things in and through which conceptual change occurred. Conceptual change is perhaps too mild a phrase—the manifest ability of living somatic cells to do various things in reaction to manipulation, such as live outside bodies, persist for endless generations, produce large volumes of virus, survive freezing, and fuse with other somatic cells, were moments of conceptual shock that troubled existing assumptions and reconfigured concepts of broad cultural salience such as individuality and immortality well beyond the confines of biological thinking. Placing tissue culture at the center of this book is thus meant to ensure close attention to change in both matter and concept, a strategy already proved productive by histories of other scientific work objects such as the fruit fly.[38]

Methodologically, the focus on practice is now well established in the history and sociology of science, and this leads inevitably to an interest in the material basis of research, particularly what Hans-Jörg Rheinberger has termed its "experimental systems," those closed-off sections of the material world that scientists employ as "machines for making the future"—that is, generating both scientific knowledge and the questions for the next experiment.[39] In the case of the life sciences, experimental systems are often composed of living objects that produce and reproduce in the laboratory environment, adding another dimension to Ian Hacking's observation that an experiment can have a life of its own, independent of theory.

The flies, mice, and cells of biological research are altered by humans environmentally or physically to do "unnatural" things, but they are not literally machines. They occupy a form of "edge habitat," where organisms with their own natural histories come into contact with and are shaped by the technological, industrial environments of human beings.[40] Living technologies such as flies, mice, and cultured cells are part of the attempt to stabilize the innate flux and variation of living things as well as to simplify and standardize the objects of research as much as is possible. In order for scientists to say that they work "the cell," they must be able to compare their experiments to experiments that they have performed at other times and that different scientists have performed in different laboratories. Genetically and physically reshaped living matter plays an infrastructural role in making biology the same over time and space.

Attention to the things people work with in experiments and to the ways they attempt to stabilize living objects such as the cell for scientific study has allowed historians and anthropologists to address the conditions under which scientific novelty is produced. Looking closely at the routine or infrastructural conditions that constantly allow the production of new things is a method for getting around having to explain all scientific developments as a "paradigm-ordered or theory-driven activity."[41] In other words, the scientist does not have to think of it first, and act on the biological thing accordingly; change can arise from the objects and practices of experimentation themselves—how cells are kept, watched, represented, manipulated, and how they react and adapt to their technical milieu. In understanding past experimentation with tissue culture—and in the current uses of cultured cells—there can be no separation of the hands-on and the intellectual reshaping of what cells are, what they can do, and what living technology is or can be.

The choice of tissue culture as the topic and object with which to trace a particular thread of material and conceptual history through twentieth-century life was not in itself enough to narrow the topic to manageability. I did not completely abandon the individual biography, the specific discovery, and the particular laboratory; but although each of these aspects gives the narrative form, the conceptual questions of the book are not organized by them.

At each point there is a bridge back from the specificity and particularity of these events and people to the more general questions of life and technology in the twentieth century. Although Materials and Methods sections of papers might seem very little like literature, reading enough of them has produced the thematic of plasticity and temporality that binds the chapters of this book together, binds the scientific entity to its more public existence, and binds the history recounted here to the present in which it was written. Medium and event are the specific materials and methods through which animals and humans have been and are reorganized as tissue culture technologies. It is through medium and event that the broader concepts and practices of plasticity and temporality unfold in the history of the cultured cell.

AUTONOMY

In the first decade of the twentieth century, the American embryologist Ross Harrison showed that he could keep fragments of amphibian embryonic tissue alive for prolonged periods of time, even though they were separated from the body they came from and kept in small glass vessels. The fragments not only continued to live but also developed as they would have had they remained part of an embryo. Undifferentiated tissues cut from a specific region of the embryo changed shape to become nerve cells with characteristic branched filaments that looked like nerve endings, even though there was no surrounding body into which to develop. Harrison's findings were greeted with both pronouncements of wonder at the immense experimental possibilities opened up by these experiments and with skepticism concerning whether these artificial conditions provided any knowledge about what cells do in a whole embryo.

Harrison was surprised by how much attention was paid to his work, commenting that it must have to do with prevalent attitudes toward organisms and their cells. He chided fellow morphologists for their obsession with "the conception of the object as it occurs in nature, the organism as a whole," writing that many scientists seemed to treat the animal body as a "sort of fetish not to be touched lest it show its displeasure by leading the offender astray."[1] The autonomy of cells in relation to the body was simultaneously a technical and a philosophical problem of early twentieth-century biology; what was an individual, after all, if a body's cells had their own lives? "Each of the elements . . . of our bodies lives without doubt a little for us, but they live above all for themselves," wrote one observer rather mournfully, commenting on the implications of tissue culture.[2]

When Alexis Carrel, a surgeon at the Rockefeller Institute for Medical Research, picked up Harrison's techniques and extended them well beyond the field of embryology, many of his colleagues, particularly in Europe, didn't believe the claims to be true. These clumps of cells floating alone in biological fluids, they said, must simply be surviving for a little while or be in the process of dying. When Carrel announced that not only could cells be extracted from the body and maintained *in vitro* but also could perhaps be kept alive indefinitely, the indignant disbelief that greeted these claims at the Academy of Medicine in Paris was clamorous enough to merit coverage in the *New York Times* under the title "Paris Doctors Ask Proof of Carrel: Skeptics Declare His Experiments on Heart Tissue Too Marvelous to Credit." Here, "well-known authorities on biological and medical questions" pronounced that photographs of the cultured tissues were not enough to show this startling claim to be true; a deputation would have to travel to the United States to see for themselves, for this finding involved "a complete upheaval in the ideas of European scientists on the properties of living tissue."[3]

Having become used to the idea that life or living matter can exist in a laboratory quite apart from an organism or a body, and in fact that it does so routinely in biological research and clinical diagnosis, we may find it hard to understand why this development was either shocking or unbelievable a century ago. Because of this familiarity, we have forgotten how to ask the basic question, How can life, once seated firmly in the interior of the body, be located in the laboratory, extracted from and stripped of the individual forms of animals or persons? Why did this change in the sense of possibility for living matter take place, and what created the conditions through which autonomous cellular life forms peculiar to the laboratory have become commonplace? This shift in practice and concept in the early twentieth century entailed pulling the "autonomous powers" of the body's "myriads of cellular units" out from under the dominating shadow of the "individuality of the organism as a whole," as Ross Harrison put it in his own reflections on the surprised attention his work received. A look at the beginnings of tissue culture in terms of the continuities and ruptures with the practices that preceded it recaptures some of the initial surprise at this manifest autonomy of the body's constitutive units and thus disturbs the assumption that an understanding of life removed from the body is natural or inevitable.

The story of Ross Harrison's nerve tissue experiments has been told before; it is widely recognized as a simultaneous ending—a definitive answer in a controversy that had been raging for decades over how nerve cells develop in the embryo[4]—and as a beginning for the technique that came to be called tissue culture.[5] As is to be expected of work that is considered simultaneously crucial and founding, Harrison's work has been the object of much historical scrutiny.[6] Although pleasing in symmetry, the description of these experiments as the ending of one thing and the incidental beginning of another quite different thing is a narrative artifact. It is exactly what they have to do with each other that has not been examined. Instead of taking Ross Harrison himself as the protagonist of this story, I have chosen to focus on the experimental event in which well-known techniques or approaches to embryonic organisms were combined to generate a new laboratory form of cellular life.[7]

When and why did the excised, cultivated somatic cell become an important experimental object? Why is the establishment of tissue culture of significance to twentieth-century biology and medicine, beyond the fact that tissue and cell culture eventually became so widely used that it is now a fundamental tool of life sciences research? That question has not been asked. It is assumed that the beginning of an important technique is of interest in itself, as a first instance and thus an intrinsically historic moment.[8] In what follows, accounts of Harrison's experiments between 1907 and 1910, their surrounding context in relation to other manipulations of living matter in embryology, physiology, and biology, and their subsequent adoption and elaboration by Alexis Carrel are directed toward unfolding the larger narrative of this book: the realization and growing exploitation of the plasticity of living matter, with interventions in plasticity tightly linked to interventions in the way biological things lived in time.

Temporality of both organisms and experiments is at the heart of this story. The specific material form of life that Harrison created—by juxtaposing extant techniques from bacteriology, embryology, and physiology—was in the first instance a technical solution to a problem of representing change over time in living biological matter, posed by an intractable debate within neurological anatomy and embryology. Harrison's cultures did not reveal something previously unseen: The nerve cell in development had been scrutinized and argued over for decades. However, in contrast to existing histological representations of the developing nerve within the embryonic body, Harrison made the nerve cell live visibly such that it could be watched, over time, outside the body.

In so doing, he created an experimental object that confounded and therefore exposed and changed assumptions as to the interiority and hiddenness of certain bodily processes and the kind of temporality in which it was possible to visualize the internal life processes of the body and the cell. Although put together with well-known techniques, the live cell growing in a hanging drop in a sterile see-through chamber was nonetheless an entirely new form of life—life *in vitro*. The temporality of experiment was continuous and abstracted from the animal body; the living cell could be watched from moment to moment. In contrast, the histological techniques that Harrison was challenging demanded a sequence of sections of the three-dimensional space of the whole body, and a sequence of bodies over time. For each developmental stage, a new individual organism was killed and sectioned to see the interior of its body at a particular point in time; the developmental sequence of embryonic development in time was both a spatial and temporal composite of these moments.[9]

Harrison's effort to observe developmental processes in a living object over time is what led to the technique of tissue culture. The extraction and maintenance of an interior life process outside the body in Harrison's cultures was not by any means the first time the animal body was taken apart. It had long been known—and put into practice—that parts of the body could survive for some time after the death of the whole. However, the very possibility of making a distinction between this kind of survival outside the body and life outside the body was due to the life manifested by Harrison's cultures. Life could be extracted from the body—detached from its connection to the *milieu intérieur*—and it could go on to not only survive temporarily but move, grow, and differentiate externally. It was this demonstration of the possibility of life outside the body in Harrison's work that led to the founding of tissue culture and the key part of its distinction from practices that came before: the survival, growth, and reproduction of the cells of complex organisms outside the body.

On this basis, Harrison's work should be understood as part of a shift in experimental practices of the early twentieth century that may be broadly characterized as a move from *in vivo* to *in vitro* forms of experimentation. The individual body of the experimental animal—one with an interior and an exterior, an opaque solid that was dark on the inside—was with the advent of tissue culture supplemented in the laboratory by an interior set of life processes that had been extracted, distributed, and persuaded to live outside the body, glass-enclosed but always in full view. *In vivo* experimentation did not stop; it was not replaced or displaced by tissue culture. Indeed it was not until about forty years later that tissue culture began to be used in the widespread manner that led to its contemporary ubiquity in the laboratories of biomedical research. However, the conceptual and practical shift signaled by its establishment in the first decade of the twentieth century indicates the appearance of a new way of thinking about, seeing, and experimenting upon the cells of complex organisms. The body was not replaced by the cell, nor reduced to it; rather, this technique substituted an artificial apparatus for the body and generated new views of the autonomy and activity of cellular life. As a result, understanding of the cell and the body as well as of their relation to one another was fundamentally altered at this time.

Growing Nerves without Embryos

In 1907, Ross Harrison grew a nerve. It grew out of a fragment of embryonic frog tissue, creeping and branching in continually changing form as its advancing end moved out of the piece of tissue and through the clot of lymphatic fluid in which it was embedded. In the paper "Observations on the Living Developing Nerve Fiber," Harrison describes this as an experiment whose "immediate object . . . was to obtain a method by which the end of a growing nerve could be brought under direct observation while alive."[10] Note the curious structure of intent here: The immediate object of the experiment was to obtain a method, though the direct object of the method was the observation of the living, growing nerve. That the method came first and the question second can be understood better in light of the fact that the question—in what manner does an embryonic nerve grow?—had been the subject of heated controversy for many decades and had at this time come to something of an impasse.

By the first decade of the twentieth century, the argument had become a three-cornered one as to whether the nerve fiber grew out in one continuous process from a single cell, arose in parts from a chain of many cells, or grew within plasmatic bridges that remained between embryonic cells after their division during development.[11] This controversy was itself part of a larger set of conflicting views about the nature of the nervous system. The so-called neuronists, beginning with Wilhelm His in 1864, held that nerve fibers were independent, free-ending structures, each originating in a single ganglion cell, and that nerve impulses were transmitted by contact between the branching endings.[12] The opposing school of thought, the reticularists, believed that all nerve fibers were connected in a continuous network or reticulum and that the growth and direction of a nerve fiber was determined by the contributions of the many cells constituting its path and substance. What was at stake conceptually in designating the nervous system as a reticulum or a syncytium of many cells was nothing less than the universal applicability of cell theory, because a reticulum would imply that nervous tissue was different from other kinds of tissues and thus that not all tissues are composed of cells. Technically, however, adherents of all points of view used the same methods to observe the same kinds of materials, and those methods were histological.[13] Arguments went on about what was or was not an artifact in different preparations, or the respective utility of different staining methods; but these were conducted within the unquestioned assumption that histology was the appropriate methodology for investigation of the problem.

Santiago Ramón y Cajal, a neuronist, and Camillo Golgi, a reticularist, were jointly awarded the Nobel Prize in 1906 for their elucidation of detailed neuroanatomy through sophisticated staining techniques. Golgi used the opportunity of the acceptance speech to argue against Cajal's neuronist views. The silver staining method that Golgi invented and Cajal refined was used to study regeneration and growth of nerve fibers from 1903 onward. For Cajal, for example, this involved the hardening of selected pieces of neural tissue in potassium bichromate and osmic acid for 24 to 48 hours, and then exposing them to the action of silver nitrate, which produced a reddish-black silver chromate precipitate in selected cells.[14]

Histological techniques of killing and solidifying tissues and then selectively staining their various cell types were at the heart of delineating the complex structures of organisms by the turn of the twentieth century.[15] One did not have to be a histologist or anatomist to view histology as a central tool of analysis; embryologists and physiologists also used these techniques to fix specimens for later analysis. The different fixatives and dyes were regarded as primary tools in visualizing normal, developmental, and pathological morphologies of tissues and cells in the body. New stains or techniques were met with excitement and regarded as marks of progress in the field. Paul Ehrlich, whose work from the 1870s onward was so central to bringing synthetic dyes into use in biological laboratories and to understanding the mechanism by which the dyes worked, rightly took great pride in these techniques. He remarked, "The dry, stained preparation is indispensable . . . To-day, we can only take the standpoint that everything that is to be seen in fresh specimens—apart from the quite unimportant rouleaux formation and the amoeboid movements—can be seen equally well, and indeed much better, in a stained preparation, and that there are several important details which are only made visible in the latter, and never in wet preparations."[16] Movement was unimportant, and staining led to an equal or better appreciation of the necessary details of the cell under observation. Furthermore, the technique of staining, by suspending the cell in time, freed the experimenter from the temporal exigencies of a living subject: "As regards the purely technical side of the question, the examination of stained dry specimen is far more convenient than that of fresh, for *it leaves us quite independent of time and place.* We can keep the dried blood, with a few precautions, for months at a time before proceeding to the microscopic investigation, and the examination of the preparation may last as long as required, and can be repeated at any time."[17] Armed with a stained preparation, the investigator was "independent of time and place" because he could examine the tissue for as long as necessary, at any time, over a period of months, repetitively.

Although Ehrlich and Lazarus were in this particular instance talking about the observation of cells for the purpose of diagnosis, fixing and staining were also regarded by many as far superior to using living tissue in anatomical investigations. Ramon y Cajal, major protagonist in the nerve debate, regarded fixing and staining as the method by which a "double invisibility" encountered in looking at living tissues could be overcome:

As if nature had determined to hide from our eyes the marvelous structure of its organization, the cell, the mysterious protagonist of life, is hidden obstinately in the double invisibility of smallness and homogeneity. Structures of formidable complexity appear under the microscope with the colourlessness, the uniformity of refractive index, and the simplicity of architecture of a mass of jelly . . . The histologist can advance in the knowledge of the tissues only by impregnating or tinting them selectively with various hues which are capable of making the cells stand out energetically from an uncoloured background. In this way, the bee-hive of the cells is revealed to us unveiled; it might be said that the swarm of transparent and invisible infusorians is transformed into a flock of painted butterflies.[18]

Unfortunately, in the context of debate over whether a nerve fiber grew via protoplasmic movement, such techniques provided exquisitely detailed specimens that were then interpreted in diametrically opposed fashions. Harrison recounts that Cajal, upon examining a rival scientist's specimens, was astonished to find that the specimens looked identical to his own.[19] Harrison argued that "the evidence for and against the two theories . . . rested upon such minute histological details that a decision to which all would subscribe was impossible of attainment."[20] Thus the controversy left embryologists and neurologists at something of a loss, for there seemed to be no way to resolve the difference of interpretation with ever more refined histological techniques; but at the same time, according to Harrison, it was a fundamental question that needed resolution: "it is obviously impossible to study intelligently the mechanics of development of the nerve paths, unless we know whether we are dealing primarily with phenomena of protoplasmic movement or with mere progressive differentiation without movement."[21]

Cellular movement, or lack thereof, was a matter of inference when using static representations, the hardened moments preserved in histological specimens. It was about inferring what was happening in the spaces between the sequential slices of preserved moments. Harrison sought to break out of the deadlock of opposing interpretations of the same material by asking the same questions but changing the material. He sought to watch the live nerve fiber as it grew; he did not wish to be "independent of time and place." The problem of visualization of translucent, tiny structures was answered not by differential staining but by a different mode of isolation and observation: the isolation of one fragment of the living embryo from the rest of the body such that every moment of its continuous development was visible. To this end, Harrison isolated pieces of embryonic frog known to give rise to nerve fibers, such as the medullary tube or the branchial ectoderm, before any visible differentiation into nerve fibers had occurred. He placed each piece into a drop of frog lymph on a cover slip, waited for the lymph to clot and hold the tissue in place, then inverted the cover slip over a hollow slide, and sealed the rim with paraffin. Fluid, glass, warmth, and asepsis were the key elements of the method; the resulting tissue, embedded in lymph and sealed in a transparent chamber made from glass slides, could be kept in an incubator and "readily observed from day to day under highly magnifying powers."[22]

This neat arrangement, called a "hanging-drop" preparation, provided a mechanical support for the cells in the fibrin present in clotted lymph. Because the drop was sealed in the chamber, the tissue was protected from bacterial infection, which would have caused it to die. The hanging-drop technique was invented in the 1880s by Robert Koch, who first grew anthrax bacilli in hanging drops of fluid taken from oxen eyes. The aim of the chamber was to isolate one type of bacilli from other microorganisms. By professional self-identification and training, Harrison was an embryologist of the *Entwicklungsmechanik* school; yet in his attempt to culture isolated nerve tissue, he drew not only on the concept of culturing microorganisms but also literally adopted the laboratory equipment, manuals, and knowledge of bacteriology by conducting the work in the laboratory of his colleague, bacteriologist Leo F. Rettger. The hanging-drop technique, which employed hollow glass slides and a drop of medium suspended from a cover slip inverted over the hollow, was by that time a common practice in bacteriology to observe living populations of bacteria and molds under the microscope.[23]

Although the use of clotted lymph and the hanging-drop technique immediately led to nerve outgrowth from the explant into the medium, the preparations were killed almost as immediately by rapid bacterial infection. Harrison thus turned to working aseptically, which involved "much tedious detail, though it offered no insuperable difficulties." It was here that the bacteriological laboratory equipment was most essential. All the glassware was flamed; the cloths and filter papers were autoclaved; and the needles, scissors, and forceps were sterilized by boiling. The embryos were washed in six successive changes of filtered water, and the frogs from which the lymph was extracted were also washed in filtered water. Harrison's description—"the making ready of the apparatus consumes so much time, and the constant attention to the details of manipulation during operations is so fatiguing, that only a small number of preparations can be made in one day"—gives us some idea of the amount of labor involved in the 211 preparations he made for this experiment.[24] As a result of Harrison's use of this aseptic technique, the preparations could be kept alive for over five weeks. Most importantly, the growing nerve fibers could be observed at all times, as they grew; Harrison did not have to section tissues after time had elapsed and infer that growth had taken place, as histological methods demanded.

What he observed was the rapid outgrowth of nerve fibers from the clump of tissue into the lymph clot. He noted the amoeboid end of the growing fiber, which branched out into filaments and underwent constant change in form, and he concluded that "these observations show beyond question that the nerve fiber develops by the outflowing of protoplasm from the central cells . . . No other cells or living structures take part in this process."[25] Harrison had included a form of control by culturing not only nerve tissue but also the germ tissue that was known to give rise to epithelium and muscle. These cells differentiated as they would be expected to within the body. Epithelial cells showed active cilial movement for weeks at a time and developed a typical cuticular border, while masses of cells taken from the myotomes differentiated into muscle fibers showing typical striations and spontaneous twitching.

Harrison's experiments on the origin of the nerve fiber had the nature of an event. He did not see or discover an object that had not been seen before; rather he made an event happen in which he claimed to witness what an object did. An editorial in *Nature*, reprinted in *Science*, reflects this difference, stating that Harrison had demonstrated the correctness of the neurone outgrowth theory in a "remarkable" way:

He has *actually seen* the fibers growing outwards in embryonic structures . . . there was no doubt that even under these artificial conditions—rendered necessary for microscopic purposes—life and growth were continuing. From the primitive nervous tissue, and from this alone, nerve fibers were observed growing and extending into the surrounding parts.[26]

Technically and representationally, Harrison was able to change the temporal and spatial parameters of observing developing nerves. Harrison sought to take the argument over the origin of the nerve fiber, as he put it, "out of the realm of inference" and to place it "upon the secure foundation of direct observation," which in practice meant out of the body and onto the microscope stage.[27] It was not a question of seeing a thing; everyone "saw" the nerve fiber in the embryonic body via dissection or histological preparations, and they observed the place and state of the nerve fibers within the sectioned tissue along with the changes at various stages of development. It was a question of making a process visible by seeing the thing change continuously over time.

Being able to see something directly did not mean that this sight was somehow less complicated or less technically mediated than histological techniques for visualizing cells. Whatever he could "actually see," Harrison still had to represent what he saw and the vast practical difference between his hanging-drop experiment and histological observation can be seen in the difficulty Harrison had in generating figures for his publications. This difficulty shows how profound a break was made with histological conventions, the type of object histology represented, and the way histology represented the passage of time in a developing organism.

Both Harrison's initial 1907 report and the later, more detailed one of 1910 on these structures are striking in their evocation of evanescence. The fibers were described as bordering on the invisible, appearing vitreous, consisting of "an almost hyaline protoplasm." The fibers, only one to three microns thick, had "remarkable" enlarged ends "from which extend numerous fine simple or branched filaments." Not only were the microscopic fibers transparent as glass but the "close observation" afforded by their transparent chamber revealed "a continual change in form, especially as regards the origin and branching of the filaments. In fact the changes are so rapid that it is difficult to draw the details accurately . . . one fiber was observed to lengthen almost 20 μ in 25 minutes."[28] The paradox involved in observing living processes was the difficulty of recording or even describing those processes. Even a drawing or a series of drawings over time presupposed the ability to capture discreet moments from a continuous process (see Figure 1). As Harrison wrote,

The character of the movement that takes place at the end of the fiber is difficult to describe. The filaments in which the fiber ends are extremely minute and colorless, showing against their colorless surroundings only by difference in refraction. The eye perceives, therefore, only with difficulty an actual movement, though when an active end is observed for five minutes it will be seen to have changed very markedly, so that in making drawings one encounters the difficulty of having the object change before the outline can be traced.[29]

Harrison reports that drawings were made with a camera lucida. This instrument allowed the observer to view the image of the specimen superimposed on the image of the paper on which he was drawing. This was not a projection of the image on to the paper, as in the camera obscura, but an arrangement of lenses and mirrors or the use of a prism placed such that the images of the drawing paper and the specimen could be viewed simultaneously. Although different camera lucidas used different prisms or mirrors,

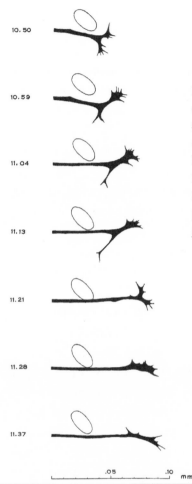

Figure 1   Harrison made these time series drawings with a camera lucida to show how a single nerve cell in culture was constantly changing as he observed its growth. From Ross Harrison, "The Outgrowth of the Nerve Fiber As a Mode of Protoplasmic Movement," *Journal of Experimental Zoology,* 1910.

in general the principle was that the observer's eye would see the separate images of the specimen and the drawing paper as one (as in stereoscopic viewing), which would give the observer the sense of seeing the image of the specimen on the blank paper.[30] Thus Harrison's description of "tracing" the outline of the nerve cell; the superimposition of the images would give to an observer the sensation of the pencil moving along the outline of the specimen. Even though Harrison could keep the growing nerve fiber constantly under observation, without having to look away and draw from memory, it changed as he drew, literally under his pencil, even though he could hardly perceive it moving.

These drawings were in fact the only permanent record left of what Harrison saw in his experiments, other than the words of his report (see Figure 2). His method, although it was designed to remedy the fundamental shortcomings of a static histology, also created an uncapturable phenomenon by histology's terms—that is, in histology's preservatives. Removed from the supports of the body and undergoing constant change, the "structures are so delicate that the mere immersion in the preserving fluid is sufficient to cause violent tearing."[31]

Appropriately, one of the combatants in the early twentieth century part of the debate, embryologist Hermann Braus, was grudgingly brought to concede the accuracy of Harrison's position only after he had personally witnessed the phenomenon by recording it cinematically. In 1905 Braus had transplanted tadpole limbs to various locations on other tadpoles, interpreting his results as supporting the protoplasmic bridge theory.[32] He and others had argued that it was possible that protoplasmic bridges remained between the cells in the piece of tissue explanted into the hanging drop; and in fact Braus continued to claim that, if one stained and dissected a culture prepared via Harrison's methods, one could find such protoplasmic bridges in the tissue. Nonetheless he conceded the point because he had observed nerve fibers grow from their "first beginning" to their full development from isolated neuroblast cells, and so he claimed for himself "the first total certainty" regarding the

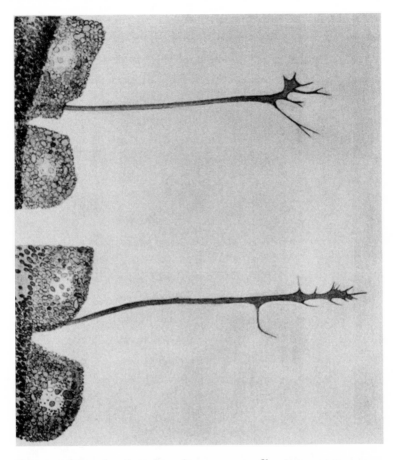

Figure 2   These drawings show the same nerve fiber in two states,
twenty minutes apart. Harrison complained of the difficulty of depicting
growth of the nerve fiber with static drawings; the time series and the
side-by-side comparison were his attempts to show change over time.
From Ross Harrison, "Embryonic Transplantation and the Development
of the Nervous System," *Harvey Lectures*, 1908.

nerve fiber: "the nerve fiber grows out of [the neuroblast] like mould produced out of an isolated spore."[33]

Although further development of the hanging-drop technique soon enabled the cultures to be histologically preserved, the turn to the new medium of cinema to capture the newly exposed "growth phenomena" (as Braus called them) nonetheless underscores the temporal break with the conventions of histology: The medium of representation had to move over time, just as the experimental object did.

The Production of Tissue Culture

Harrison's work between 1907 and 1910 was greeted as proof of the single origin of the nerve fiber with a mixture of enthusiasm and grudging confirmation. It also evoked dismissive responses, especially from the opponents of the outgrowth theory, on the grounds that the culture conditions were manifestly artificial and thus not informative about normal development in the embryonic body.[34] The greatest enthusiasm, however, came from scientists not involved at all in the debate but nonetheless fascinated by the life manifested by the nerve cultures.

In 1908 Harrison was invited to lecture to the Harvey Society, a predominantly medical association that met monthly in New York City to hear lectures on issues of biology, public health, and medicine. Here he described the method, his results, and the role of these experiments in the controversy over the origin of the nerve fiber.[35] Shortly after, he began receiving letters expressing interest not so much in the nerve fiber controversy as in the method he had described. Dr. W. G. MacCallum, a pathologist at Columbia, wrote to Ross Harrison in 1909, saying "I am anxious to have one of my assistants Dr. Lambert learn from you some of your methods which you have applied to the study of the growth of tissue such as nerve." MacCallum's interest was quite specific; he wanted to see whether the method could be used to study tumor growth. *"It may seem fantastic* but I wondered whether it might not be possible that in suitable surroundings the rapidly growing tumor cells of mouse cancer might be induced to grow apart from the living mouse."[36] Lambert met with no success in trying to carry out MacCallum's "fantastic" idea of growing mouse tumor cells without a mouse, despite learning Harrison's technique directly from him.

At the same time, Warren and Margaret Reed Lewis at Johns Hopkins were attempting to grow fragments of tissue using Harrison's new method but substituting mammalian blood plasma for frog lymph. Warren Lewis had worked with Harrison at Johns Hopkins, and Margaret Lewis had spent the previous year in Germany, where she had achieved what she thought was growth of cells outside the body. "In 1908, working under Dr. Max Hartman at the Kgl. Institut für Infektionskrankenheit, one of us (MRL) found that bone-marrow from the guinea-pig formed a membrane-like growth with mitotic figures on the surface of nutrient agar. This agar was a modification of one used at the Institute for the cultivation of amoeba."[37] In light of Harrison's results, Margaret Lewis's observations seemed more significant, and they thought to try again to induce cells to divide in nutrient medium. These experiments were not successful, and the Lewises gave it up for the time being.

Dr. Alexis Carrel of the Rockefeller Institute for Medical Research was in the audience for Harrison's lecture to the Harvey Society in 1908, with his own troubles on his mind. Head of a unit called Experimental Surgery at the Rockefeller, Carrel had for many years been interested in cell regeneration in the context of wound healing or, in his own terminology, the "laws of redintegration of the tissues of mammals."[38] A prolific transplant surgeon, his successes were being frustratingly confounded by the inexplicable deaths of animals into which organs had been transplanted. Although he could take organs and limbs out, detach them completely from the animal, sew them back into the same animal, and see them regain function, transplants between animals were not working. In retrospect, Carrel reported listening to Harrison, not with any interest in questions of embryology but with the thought of finding a method "permitting the cultivation, with constant positive results, of mammalian tissues outside of the body"; this led him to "acquire and improve the technique of tissue cultivation."[39]

In 1910, with encouragement from Simon Flexner, Carrel's new assistant Montrose Burrows wrote to Harrison:

At Dr. Flexner's suggestion I am writing to ask if you would be willing to have me come and work with you for a month or so this spring. The work I have been doing here with Dr. Carrel has consisted in an attempt to explain the mechanism of nerve degeneration and secondly an attempt to stimulate the regenerative process of cut nerves. Dr. Carrel has suggested that we might best attack the problem on lower animals or better on artificially grown nerves by your method.[40]

Burrows met with much more success than Lambert or the Lewises. Montrose Burrows had graduated from Johns Hopkins University in 1909 with a medical degree and gone to work with Carrel as a junior fellow at the Rockefeller Institute for Medical Research.[41] In the spring of 1910, he spent several months in Harrison's laboratory at Yale in order to "adapt, if possible, his method to the investigation of the growth of the tissues of warm-blooded adult animals."[42] Burrows repeated Harrison's experiments but substituted frog blood plasma for lymph, with positive results. The use of plasma, which clotted like lymph but was much easier to obtain in quantity and was more homogenous in quality, made the preparations less laborious to construct and more reliable. He then substituted the embryonic chicken for the embryonic frog and chicken plasma for frog plasma. Overall, the hanging-drop/hollow slide arrangement was the same, except that the tissues had to be kept warm because they were taken from a warm-blooded animal. The nature of the tissue was different: Frog embryo cells are filled with nutrients in the form of yolk granules but chick embryos are nourished from an extracellular yolk via a vascular system. Excising pieces from the vascular system thus meant that the surrounding medium had to be a nutritive medium for the cells as well as a support. Thus the bodily functions of heat and feeding were added to the apparatus surrounding the tissues.

Burrows repeated and confirmed Harrison's results and at the same time settled his own controversy: the myogenic origin of the heartbeat. When Burrows cultured embryonic chicken heart cells, the fragment of live tissue embedded in the drop of plasma continued to beat. Cells began to wander outward from the fragment into the surrounding medium, and Burrows could observe single, isolated cells themselves pulsing rhythmically. The physiological controversy that this addressed was whether the heartbeat was caused by stimulation of heart muscle by the nervous system (the neurogenic version of events) or the heartbeat arose from within the heart muscle itself (the myogenic argument). This controversy bears some parallels to that over the origin of the nerve fiber. Both conflicted over the level of interconnectedness versus autonomy of the different elements of the body, both reached a certain point of intransigence mired in the methodological difficulty of isolating the supposedly interconnected or supposedly autonomous elements from each other while still maintaining a living experimental subject.[43] The method of tissue culture allowed the clear separation of heart tissue from any possible nervous elements, and the spectacle of the single cell pulsing in a "perfectly normal rhythm and with a rate and force similar to the heart in the embryo" was certainly strong evidence for cellular as well as muscular autonomy or, as Burrows phrased it, "the automaticity of the heart muscle cell."[44]

On October 15, 1910, Carrel and Burrows published the first of four closely spaced publications in the *Journal of the American Medical Association* on tissue culture.[45] From the very beginning, they cited Harrison's "beautiful work" as the "starting point" of their research; but in adapting the technique, they changed it profoundly. On the surface the technique was the same—a fragment of living tissue was placed in a hanging-drop preparation and maintained for a period of time. However, a number of substitutions meant that the method was fundamentally changed. Carrel and Burrows substituted adult tissues for embryonic ones, mammalian for amphibian, and blood plasma for lymph. Perhaps the most profound change was the act of making secondary cultures by taking fragments of the original culture and placing them in new plasma. They called this "reactivation and cultivation in series." This changed the timescale of the experiment, from one-time preparations that lived for days or weeks, to a potentially endless series of preparations each made from another. The relation of the *in vitro* tissue to the body of the animal was therefore also different—the researcher did not return to the body each time he or she made a culture, but could make cultures from other cultures already living outside the body.

Only two weeks later, Carrel and Burrows reported on the "Cultivation of Sarcoma Outside of the Body."[46] This research employed a further substitution—cancerous tissue for normal tissue. They used a sarcoma that Carrel's colleague Peyton Rous had been studying in chickens. They reported that Rous had been propagating the sarcoma "from generation to generation for more than a year," which means that Rous had been transplanting bits of the tumor from one chicken to another for a year, growing new tumors in new chickens, and thus in a sense doing serial cultivation of the tumor tissue in the bodies of the chickens. It is possible that Rous's work suggested the serial cultivation of the tissue outside the body, although the inspiration could have come from bacteriologists, who commonly made new microbial cultures from old ones. Carrel and Burrows observed that cancerous tissues started to grow much more quickly than normal tissues, which usually stayed fairly inert in the first hours or even days in culture, and only after some time started to move and divide.

The third substitution was to replace chicken sarcoma with human sarcoma, attained from a patient who had a tumor removed from her leg. However, human tissues proved extremely difficult to culture and the cells lived for only a few days in culture. Using human tissue was clearly possible in principle but very challenging in practice, which made for an extremely brief reporting of results. The fourth paper varied both the tissue type and the medium it was grown in. The growth of normal and sarcomatous chicken tissue was compared in normal blood plasma and plasma taken from chickens with tumors. Carrel and Burrows found that the "plasma of a sarcomatous animal acquires the property of inhibiting the growth of sarcoma taken from another animal," concluding that this must be due to "substances produced by the organism as a reaction against the tumor."[47] Furthermore, embryonic spleen tissue grew faster in sarcomatous than in normal plasma.

These four papers, which set out the possibility for the medical and biological public of growing all kinds of tissues outside the body, were very different from Harrison's publications on the developing nerve. There was a cumulative nature to Carrel and Burrows' substitutions—of blood for lymph, of mammalian tissue for amphibian, of serial cultivation for one-off cultures—that left the method in a very different place than they had initially picked it up. With this reconfigured system, they had no specific questions to answer such as Harrison's pointed investigations of the growth of the nerve fiber. Instead, the very possibility afforded by the general method of tissue culture was their subject and conclusion. The papers contained extensive descriptions of what the cells looked like and a few results, such as the comparison of the effect of different media on different tissues; but for the most part their emphasis was on the new experimental possibilities of tissue culture. First, it was possible to grow all kinds of embryonic and adult tissues, normal and pathological—and importantly, also human tissues—outside the body: "The main results of these observations can be summarized in a few words: Adult tissues and organs of mammals can be cultivated outside the animal body. The cultivation of normal cells would appear to be no more difficult than the cultivation of many microbes."[48]

Second, this life and its growth were "luxuriant," with the cells growing in states of "wild vegetation, which lasted as long as the plasmatic medium was in good condition."[49] This growth could be "reactivated" at will by simply making a second culture from the first. Third, all of these tissues could be grown in isolation "under known conditions," conditions that could be varied at will. Not only would this give access to information about normal tissues but also it promised to be important in the analysis of pathogenic processes: "it may render possible the cultivation of certain microorganisms in conjunction with living tissue cells . . . Then it will be of great value in the study of the problem of cancer."[50]

Finally, the proliferation of the cells grown in this way could be constantly observed over time, and phenomena such as cell division "directly seen": "The purpose of the present article is merely to show that all the details of the living cells can be observed at every instant of their evolution."[51] To drive this point home, they went into detail describing the appearance of a single cell in culture on October 29, 1910, at 9:00 AM, 9:03 ("it became slightly oval"), 9:06 ("it was more oblong"), 9:18, 9:20 ("there was great activity among the protoplasmic granules of the posterior end"), 9:22, 9:25 ("the tail was longer. The cell also had increased very much in size."), 9:30, and finally 9:45, when "the cell had assumed the same appearance as before 9 o'clock." This observation shows how accurately the living cell can be studied in a culture."[52]

In the space of four months, Carrel and Burrows had transformed Harrison's method for the short-term growth of living embryonic tissue into a generalized method for the cultivation of all kinds of tissues, embryonic and adult, amphibian, mammalian and human, normal and pathological. In four publications that together totaled just seven pages in the *Journal of the American Medical Association* (results that were simultaneously published in French in the *Comptes Rendus des Séances de la Société de Biologie*), they outlined the possibility of serial cultivation of these tissues outside the body in media of known composition as well as their use in the analysis of cancer and other pathogenic phenomena afforded by the ability to observe the cells "at every instant of their evolution."

Thus from 1910 on, Carrel and Burrows appropriated and altered Harrison's technique, coined the term "tissue culture," defined it, announced it to the world, and set it on its way to becoming a generally applicable tool of experimental biology with much further reaching uses than either nerve fiber research or embryology. Harrison had used embryonic amphibian tissue, but Carrel and Burrows used adult and cancerous mammalian tissues, grown in serum rather than lymph; and they introduced the idea of continuous culture—making new cultures from old ones, without returning to the body of the animal for new cultures.

The term "tissue culture" itself, which quickly came to denote both the method and the field of knowledge created by it, was first formally defined in 1911 by Carrel and Burrows as "a plasmatic medium inoculated with small fragments of living tissue"—importantly, the definition included both the attributes of growth *and* reproduction.[53] New populations of cells could be raised by taking a fragment of an old culture and putting it in new medium, which meant an ongoing cultivation of somatic animal cells without further recourse to the animal body—something that Harrison had neither tried nor mentioned as a possibility. Where Harrison's interest lay in observing differentiation and movement, Carrel's was directed toward observing "life" and its essential characteristics—growth and reproduction—outside the body.

Tissue culture was quickly incorporated into cancer research, immunology, virology, and other cellular research programs. Carrel and Burrows always recognized Harrison's work as the beginning of their own (albeit in a slightly backhanded manner); and a short two to three years after the initial publication of his results, Harrison became an instant historical figure: a founder, always invited to come and speak to the history of the technique, even before it had much history to speak of. Ironically enough, he himself had stopped using the technique in any major way by 1914 and had moved on to other things; perhaps this distance just accentuated the perception of him as a representative of the "history" of the new technique.

As Carrel and Burrows saw it, the main experimental result of Harrison's work was a sense of possibility. The function of Harrison's experiments was to unequivocally demonstrate the feasibility of growing tissues outside the body; and the similar responses of MacCallum and the Lewises indicate that Carrel and Burrows were not alone in their perception of a new experimental opportunity. They wrote, in recognizing Harrison's work: "These experiments demonstrated that the nerve fibers are really an outgrowth from a central neurone. But they demonstrated also a *very much more important* fact, the possibility of growing tissues outside the body."[54] In what situations are possibilities regarded as facts? Usually those in which the impossibility of something has heretofore thought to be true. To them, an answer to the nerve outgrowth controversy was all very well, but there were larger "facts" at stake: It was a fact, given Harrison's results, that tissues could grow outside the body.

Taking the Animal Body Apart

Nobody ever pretended that tissue culture arose without precedent from Harrison's brain or laboratory, least of all Harrison himself, who expressed his surprise that everyone else seemed so surprised by the technique:

> . . . this method of isolation of cells or pieces of tissue is but the application of the method of the physiologist when an organ is isolated in order to find out its function, or that of the experimental embryologist when he isolates the blastomere of the segmenting egg to determine its developmental potencies. Technically, the method is an adaptation of one that has been for years familiar to the bacteriologist—the hanging

drop culture. Any originality, therefore, that may be claimed for this work is due to a combination of ideas rather than to the introduction of any particularly new device.[55]

There is no shortage of convincing lines of continuity to be drawn between Harrison's nerve culture experiments and his own earlier embryological investigations and work with heteroplastic grafting.[56] Embryologists had already been observing isolated fragments of living embryos for some decades and doing cell lineage studies on cellular movements within early embryos.[57] Historian Frederick Churchill has described the period 1885 to 1901 as a time in which "experimentalists and anatomists turned to their scalpels, scissors and heated needles." With these instruments, they "shook, ligated and compressed . . . all with the intent of understanding the 'meets and bounds' of regeneration."[58] In a letter from the marine biological station at Naples, Harrison himself reported in 1896: "Nearly everybody is shaking eggs of some animal or another—it looks as if one were behind the times not to do it."[59] Techniques of excision, transplantation, and disruption were applied on a very wide scale to test the conceptual issues of the regenerative and developmental capacities of parts of the embryo.

Transplantation and isolation of organs was also carried out on animals and humans by surgeons and physiologists. Historian Anne Marie Moulin writes that the motives for transplantation at the end of the nineteenth century had a double nature that corresponded respectively to surgical and biological undertakings: the logic of the surgeons was substitutive, that of the biologists provocative, directed toward testing the limits of an interpretation of nature and living matter.[60] Further, she proposes that the experiences of transplantation in both surgery and biology served to demonstrate the plasticity of living beings, delimited at the beginning of the twenti-

eth century by apparently insuperable limits on the ability to transplant tissues or organs between species. These limits constituted a
core practical and conceptual definition of individuality in biology.

There are many interesting connections to be drawn between
tissue culture and this cornucopia of transplantation and regeneration experiments of the latter half of the nineteenth century, which
Jane Oppenheimer characterized as a period of "taking things apart
and putting them back together."[61] It is not difficult to see Harrison's work and tissue culture in light of previous attempts to culture tissues outside the body in embryology, pathology, or tumor
transplantation.[62] Parallels have also been drawn between Harrison's nerve cultures and earlier work on body cells that were more
easily separable, such as blood, egg, or sperm cells,[63] or with the
maintenance of life of organs surgically removed from the body
via perfusion.[64]

In sum, Harrison's work took place within an intellectual and
practical landscape of widespread probing of the bounds of bodily
individuality, partability, and plasticity. The cellular nature of tissues and organisms, although still debated (and particularly controversial in the case of the nervous system), was widely accepted. Microorganisms such as bacteria and protozoa were cultivated in
nutrient media, and scientists from cancer pathologists to bota-
nists had been influenced by bacteriology's productivity and success with bacterial culture to try to grow cancer cells and plant cells
in nutritive medium.[65] In fact, one could potentially bury Harrison's
experiments or the development of tissue culture in these various
contexts in such a way as to make them disappear as events of note
at all.

This would be a mistake, however. The paradox of the production of something new out of well and widely known extant practices and concepts—the question of the event of novelty even in

the light of rich contextual connections—is relevant to much of the history of science, and it is central to understanding this particular case.[66] Making connections and discovering precursors does nothing to explain why tissue culture appeared to Harrison's contemporaries as something startlingly new, perhaps even unbelievable—an idea that needed to be tempered with the admission that to try such an experiment "may seem fantastic," as MacCallum's letter to Harrison put it. One excited assessment of 1911 in the *Journal of the American Medical Association* compared the advent of tissue culture to the synthesis of urea in the nineteenth century:

> Wöhler in 1828, by synthesizing urea, performed what up to that time was considered to be impossible, namely, the artificial production of an organic substance from mineral matter. Since then up to the present time there have been few advances in biology which seem so startling as do those which have been made in recent years, resulting in the establishment of a technic whereby animal tissues may be made not only to live, but also actually to grow outside of the body.[67]

In this analogy with Wöhler's synthesis of urea, it is evident that tissue culture had a twofold capacity to surprise: first, the performance of something previously assumed to be impossible, and second the artificial nature of achieving organic substance via inorganic means. Of course, in the case of tissue culture this was not generation *de novo* from inorganic materials. The comparison should nonetheless be noted because it shows that this process of inducing growth outside the body appeared artificial—and this artificiality was part of the sense of novelty surrounding the development of tissue culture.

Others simply disbelieved that such growth and reproduction of somatic cells *in vitro* was possible at all. Albert Oppel, of Halle, who was later sufficiently moved by the technique to write one of the first books on tissue culture in 1914, *Gewebekulturen und Gewebepflege im Explantat* (with an introduction by Paul Ehrlich), declared that at first he dismissed the reports of cells growing outside the body as sensationalist fairy tales: "At first I didn't believe them any more than when I was assured it would be possible, through the injection of organ extract, to double the height of a full grown man in his 30s or 40s. I took it for no more than a fairy tale, such as I heard about my own work as told in America not long before—that through merging two frog's eggs of ordinary size, I had succeeded in producing from them a two foot long frog."[68]

Even scientists who have been called (in retrospect) the forerunners of tissue culture at the time declared indignant disbelief in any such possibility of life outside the body. Justin Jolly, a histologist at the Collège de France, in 1903 kept extracted Triton blood cells alive for up to a month, and he did observe cell division in some of them, which led Georges Canguilhem to claim that Jolly had at that point invented the *in vitro* culture of explanted cells.[69] However, in 1903, Jolly himself saw these experiments to be in full continuity with previous efforts to maintain the survival of excised hearts of cold- and warm-blooded animals as well as to cultivate egg, sperm, and blood cells outside the body. He wrote that his observations on blood cells served as confirmation of the earlier work of Recklinghausen and Ranvier in the 1870s, who had observed amoeboid movement of leucocytes in drawn blood. As to the direct observation of cell division for an extended period of time after extraction from the body, Jolly simply noted that this persisted for much longer than had previously been supposed possible. Because the cells are living, "nothing is easier than to follow, eye at the microscope, the successive phases of cell division."[70]

Despite the manifest similarities that one can trace between

Jolly's 1903 work and tissue culture—maintaining living cells outside the body, observing their division—Jolly was incensed by Carrel and Burrow's reports and declared heatedly that it was an "abuse of language" to call the results "cultures." In his opinion, the phenomena being observed therein were clearly those of necrosis:

> At present, Carrel and Burrows seem to have demonstrated nothing other than the phenomenon of survival. Certain of their descriptions seem to actually be related to the phenomenon of necrosis. In certain tissues, *in vitro,* it is well known that cellular multiplication seems to be able to take place for some time; but between that final effort of some cells and a "culture," with continuous and progressive development, there is an abyss, that may one day be filled up. For the moment, it is an abuse of language to give the name of "cultures" to the results obtained.[71]

Thus we are faced with a paradox of something new assembled out of nothing new. Why, given the context, should tissue culture have seemed at first like a sensational rumor, or an "abuse of language"? Why, for those who *did* believe the results, was it so startling nonetheless, promising to suddenly "[lay] bare practically a whole new field for experimental attack on many of the fundamental problems in biology and medical science"?[72]

Although Harrison combined extant techniques to build his own, the object that he created disturbed and exposed assumptions about the interiority, limited accessibility, and invisibility of certain bodily processes—assumptions that were embedded even within the very techniques that he drew upon. The answer to how something startlingly new could arise from a juxtaposition of well-known practices and theories lies not in the disruption of the whole/part relationship implicit in tissue culture but in the disturbance of a different set of boundaries or categories in experimental practices using the bodies of animals: interior and exterior. I made the claim at the outset that Harrison's experiments and their role in the development of tissue culture consolidated an important shift in experimental practice concerning animal bodies. This shift from *in vivo* to *in vitro* entailed the knotting together of practices of visualization, isolation, and cultivation of tissues that involved not just a move from animal body to glass vessel but also a change in concept and practice that can be usefully understood as a making of what had been inside and had constituted a powerful physiological notion of interiority and invisibility live outside, visibly and autonomously.

According to Harrison, despite all of the activity described above in experimental physiology, pathology, bacteriology, and embryology, the organism as a whole still had an overwhelmingly powerful hold on his contemporaries' views of animal bodies:

> it seems rather surprising that recent work upon the survival of small pieces of tissue, and their growth and differentiation outside of the parent body, should have attracted so much attention, but we can account for it by the way the individuality of the organism as a whole overshadows in our minds the less obvious fact that each one of us may be resolved into myriads of cellular units with some definite structure and with autonomous powers.[73]

I think this diagnosis is for the most part correct, but we can give more specificity to the exact manner in which the "individuality of the organism as a whole" overshadowed the possibility of the autonomy of its parts. It was not a hesitation to cut the body into parts. There was clearly a great deal of activity all over Europe and America—in different communities of practitioners from surgeons to embryologists—that involved taking the animal body apart, keeping its parts alive for limited periods of time, rearranging the body or destroying part of it, and even sewing several bodies together to see what would happen. These practices do not indicate an inhibiting respect for the whole; rather, what they show is the dominance of the idea of the necessity of the body for the maintenance of the life of the body's parts. The assumption that pieces isolated from the body were in fact dying was a deeply held assumption: Though these pieces might be kept isolated for a period of time, they were fully expected to perish relatively rapidly.

Thus Jolly easily admitted the well-known fact that tissues in culture made a "final effort" at survival, characterized by some cell division after excision from the body and placement *in vitro*. Where Carrel saw ongoing life outside the body, Jolly saw temporary survival once tissues were detached from the body. In fact, all kinds of experiments in embryology and physiology depended on this temporary survival of body parts or organs. Rather than describe this as "living," Jolly understood it as "dying"—had not Bichat famously observed a century before that the body does not die all at once but in parts? Only reattachment to the body (a body), in whatever fashion, could stave off this death. As Jane Oppenheimer pointed out, bodies did not necessarily have to be put back together in the same way as they were taken apart.[74] However, the access of a body part to the functions of the body's interior fluids and processes was held to be essential to the continued life of that part.

A central tenet of nineteenth-century physiology was Claude Bernard's formulation of the *milieu intérieur*.[75] The body's blood vessels, nerves, and respiratory organs produced this milieu for the individual cells constituting the "organic edifice" of the body; the complex structures serving cells and tissues were "like the factories . . . in an advanced society which provides [its members] with the means of clothing, heating, feeding, and lighting themselves."[76] Bernard himself had proposed that if the conditions of the *milieu intérieur* could be reproduced around them, cells could live apart from the body: "If it were possible to reproduce at every moment an environment which neighbouring parts continually create around a given elementary organism, such an organism could live free in the same way as in society."

He went on, however, to immediately state the impossibility of such a replication of the inner fluids of tissues and organs. Cells, or "elementary organisms" as he also called them, could live their lives only at their proper place in the morphological plan:

> In the present state of our knowledge it would be impossible to reproduce artificially the internal milieu, in which each cell could live. The conditions of this milieu are so delicate that they elude us. They exist only in their natural place assigned to each element by the morphological plan. Elementary organisms do not find those conditions except in their proper place, at their post; if we transported them elsewhere, displaced them, or worse still, extracted them out of the organism, we would modify their milieu, change their life or even make this life impossible.[77]

François Jacob phrases this formulation of the sustenance of life through maintenance of a *milieu intérieur* as the idea that "higher animals literally live within themselves."[78] With the cultivation of tissues, all the processes of that inner life, of growth, division, me-

tabolism, movement, senescence, and infection, happened outside
the confines of the body. In other words, animals apparently could
also live without themselves, and this came as a surprise.

Bernard's formulation of the *milieu intérieur* was influential, particularly in France. However, it is neither necessary nor practical in
terms of historical evidence to claim that this particular formulation was the grounds for all assumptions about the equation of ongoing life with wholeness and interiority of organisms. Rather,
reading Bernard's elaboration of the *milieu intérieur* helps illumi-
nate the more and less explicit assumptions about interiority evi-
denced by the way experiments were done across Europe and
America. These experiments depended on change occurring inside
the body, change that could only be observed by cutting open that
body. Putting internal processes on view under glass broke with
these practices. This is the essential point of difference between
Harrison's experiment and those that he drew upon, and his contemporaries recognized this difference, particularly when it was ex-
aggerated by Carrel's rather more spectacular pronouncements.

This important point requires some elaboration. In experimental
embryology, interventions were made in the developing embryo of
various kinds; pieces were transplanted into other parts of the
body or slipped under the skin, parts of the embryo were destroyed, and halves of different species were bound together with
wire. These interventions were made, the interior of the body was
disrupted, and the body was closed up again and left for a period of
time. Later the body was opened again by dissection or sectioning
to see the results of the experimental disruption. The same happened with tumor transplants. The body of the animal bearing
the tumor was opened, a piece of the malignancy was removed,
the body of another animal was opened, the tissue fragment was
placed inside, and then the body was sewn up and left for some-

thing to happen to the transplant. In parabiosis, two or three animals were opened, and then the open edges were sewn together. After a period of time or, frequently, after the death of the animals, they were sectioned to see what had happened inside their bodies.

Grafting or transplanting pieces of tissue to a part of the body where they would not naturally occur was a form of isolation in that the piece was isolated from its usual location in the body and subjected to the influence of a foreign location. However, it was also a way of keeping those fragments alive, by reinserting them within the body, surrounded by its warmth, nutrition, and structural support. Take, for example, an experiment Harrison performed prior to moving on to the hanging-drop technique: He removed the entire spinal cord from a frog embryo and then transplanted a bit of medullary cord from another embryo under the skin of the abdominal wall of the first embryo. After several days, the only nerves he found in the specimen originated in the transplanted tissue, but he did not observe their growth directly. The intervention was made, the wound was closed, a few days were allowed to pass, and then the specimen was dissected or sectioned to see what had happened. When attempting to isolate the transplanted piece from other tissues, Harrison first tried to grow it in saline solution and, when nothing happened, he turned back to the embryonic body: "Later a more natural environment for the isolated tissue was sought in the ventricles of the brain and in the pharynx of young embryos."[79] Because the fragments could not be observed directly, the embryos were killed after two to seven days and examined in serial section.

In all of these experimental interventions, the interior of the body was understood to be the place of growth or change. Reinserting or reattaching the excised tissue to the milieu inside the body was seen as a necessary condition of the intervention; otherwise, it would not live long enough for an informative developmental or pathological change to occur. The main method for observing the result of the intervention was histological. Having left the tissue inside the body for some period of time to undergo this growth or change, the researcher would open up the animal and then section and stain the tissue or organ in question and observe it histologically. That is, the animal and the tissue in question had to be stopped in time—killed—in order to be observed. And these histological representations were an important part of the published results of such experiments.

Certainly there was also something at stake in the contrast of morphological methods that observed the body as it was and experimental methods that undertook various forms of intervention, and Harrison's work has been much discussed as an example of this transformation in the American context.[80] However, both morphologists and anatomists used histological methods to observe their subjects in order to build a picture of what was happening inside the body, whether or not experimental intervention was part of the program (as in the disagreement between Harrison and Braus, both of whom were doing limb transplantations). Thus Harrison's experiments with growing nerve fibers were orthogonal temporally to both experimental and morphological approaches.

Harrison and Carrel's colleagues, in greeting tissue culture as "a complete upheaval in the ideas of European scientists on the properties of living tissue," explicitly coupled that upheaval with a break from widespread practices of transplantation: "This was entirely different from merely finding a method of inducing fragments of flesh to grow on another living organism."[81] It was the autonomy of the cells that surprised them—cells could live without being reattached to a living body, without being inside a body.

Herein lies the important shift from *in vivo* to *in vitro*. In observing the living subject over time and the living process as it happened, the assumptions embedded in histological practice were confounded. One did not have to kill the animal or the tissue to observe the development course or experimental alteration of internal structures and processes. Internal processes could be placed on the exterior, and watched, given the appropriate technical substitution of particular functions of the body: asepsis, fluid, structural support, warmth. In substituting a glass enclosure and a drop of lymph for the body, something opaque was replaced by something transparent, and the enclosure did not have to be opened or halted in order to observe what was going on inside it. In not just taking the animal body apart, but leaving it apart, cellular life that was autonomous, external, and dynamic came into being for biology.

IMMORTALITY

Immortality is a concept with a heritage much longer than that of tissue culture; but in the twentieth century, the two became inextricably entwined when Alexis Carrel claimed that cultured cells could be inducted into a state of immortality using the right techniques. By describing immortality as something that could be investigated empirically using a controlled system of cells growing in a nutrient medium and a glass vessel designed by the scientist, Carrel framed the concept as a tangible object of inquiry in the field of cell biology. New scientific objects such as immortal cells do not appear out of nowhere: They may exist as entities "picked out by colloquial nouns, long before they become scientific objects."[1] Lorraine Daston observes that it is not "the absolute novelty of the thing but rather the heightened, almost obsessive attention paid to the objects and the dramatic shift in the relevant vocabulary" that occurs when objects come under energetic scientific scrutiny.[2] Such things are thus endowed by scientific attention with new forms of representation, elaborate theories, and, in turn, altered cultural significance. Immortality has, in the twentieth century, undergone several such cycles of capture from colloquial language into scientific practice and back into an altered state of cultural salience, beginning with Alexis Carrel's use of the term as a technical descriptor for tissue culture and repeating, in altered form, with the establishment of HeLa, the first widely used human cell line (see Chapters 3 and 4).

Calling cultures immortal only strengthened the perception of cells' potential autonomy from the body. The spatial reorganization of cells, releasing them from the bounds of the original organism, also seemed to free cells of the body's limited life span. The thorough alienation of cells from their originating bodies was sealed when the excised pieces of tissue lived longer than the organism itself. One reason why immortality became so thing-like, with the solidity of an empirically measurable existence, was because of its early association with one object in particular: the immortal chicken heart.

Alexis Carrel chose to use a culture of embryonic chicken heart tissue to demonstrate immortality. Soon after learning and adopting Harrison's culture technique, Montrose Burrows had shown that the heartbeat originated in the heart muscle cells. He proved that even isolated cells that had wandered away from the original explant into the surrounding medium would pulse, despite their detached solitude. Carrel then took advantage of the manifest liveliness and frank uncanniness of isolated heart muscle tissue beating to underscore his claims of "reactivation" and "rejuvenation." The cessation and subsequent artificial restoration of the pulsations of the heart muscle cells carried, for scientific and popular observers, all the connotations of the heart as the seat and sign of life and the cessation of the heartbeat as the sign of death (it would be some time before brain death was used for this purpose).

In Chapter 1, I showed how autonomous cells removed from the body came to live in the laboratory; and in this chapter, I detail the establishment of their permanent existence *in vitro*. Immortality or permanent life was not just an abstract idea imputed to cells. It was a set of specific technical interventions in physical matter that resulted in a material form for this concept of biological infinitude. Because Alexis Carrel was so central to the development of techniques for manipulating the physical environment of excised living matter, I concentrate my attention here on his laboratory at the Rockefeller Institute for Medical Research, although others were also working on tissue culture in the same period. Carrel's focus on duration (which he understood via the philosophical writings of Henri Bergson) shaped tissue culture work in this era. For better or worse, Carrel left an enduring legacy both in terms of technical form (glassware, lab layout, protocol) and in terms of expectations and emphasis—tissue culture as a method to study the cell as a dynamic, temporal being.

The contrast between Ross Harrison and Alexis Carrel is striking. Harrison was an experimental embryologist seeking material ground to test two opposing theories of nerve growth and cellular autonomy; Carrel was a surgeon, with a much stronger tendency to tinker with tissues in an open-ended way—to see how far one could push them and what would happen when one did—than to experiment in a highly controlled, hypothesis-driven way. For this and other transgressions, Carrel has earned the scorn of later generations of scientists (particularly those who worked with tissue culture). His work is well worth understanding nonetheless. Historian of biology Philip Pauly has written that the work of Carrel's colleague at the Rockefeller Institute, Jacques Loeb, is significant because it helped introduce an "engineering ideal" into American biology: Do things with living matter first and worry about explanations of mechanism later.[3] Sometimes style, approach, and a sense of possibility concerning living matter are as—or more—important than identifiable "discoveries." Similarly, in the case of

Carrel, the ability to do things with cells, to control their temporality by intervening drastically in the structural and mechanical conditions of their physical existence, is an important part of the genealogy of contemporary biology and biotechnology, regardless of his personality or eccentricity.

There is much to say about Carrel, a controversial figure with a long career bracketed on one end by spectacular transplant surgeries and a Nobel Prize and on the other by eugenic writings directed at a popular audience and an ignominious death in 1944 in occupied France.[4] I will limit my discussion to his work with tissue culture and his experimentation with biological time. Immortality as a scientific object emerged progressively over the 1910 to 1914 period; the "reactivation" of tissues became "rejuvenation," then "regeneration," and then "permanent life." Finally came "immortality," which was used as a technical term to designate the state of certain cells. Carrel's work on the "permanent life" of tissues outside the body was interrupted by the outbreak of war in his native France; but even his stint serving on the French front tending to wounded troops was dedicated to investigations of biological time in the progress of wound healing—experiences he brought back and applied directly to the practice of tissue culture. After 1918, Carrel developed a tissue culture apparatus, including glassware and microcinematography, to study life *in vitro;* and he began to explicitly theorize this cellular life in terms of a Bergsonian philosophy of duration. I describe the techniques, glassware, and laboratory equipment of tissue culture as a kind of operationalized philosophy of biological time. These techniques, although not necessarily maintained along with their accompanying philosophical language, were taken up by other scientists interested in the problem of establishing long-term, perhaps indefinite, populations of cells as experimental objects. Finally, I cover the very public life of the immortal tissue cultures developed by Carrel as a way of showing what happened to immortality in broader cultural terms, after it had come under intense scientific scrutiny and became attached to scientific objects such as cultured tissues.

Permanent Life

Harrison showed the potential for making internal processes external and thus visible over time. Carrel and his assistant Montrose Burrows made a series of substitutions in Harrison's method and ended up with a method at whose core was "life" outside the body, which was defined not as survival but as growth and reproduction of tissues. Carrel and Burrows substituted adult tissues for Harrison's embryonic ones. They introduced continuous life to the culture by making new cultures out of old ones instead of out of organisms. An experiment was no longer bounded temporally by a finite survival period, an intact body was no longer the only source for living cells, and an organism was no longer the only location for the reproduction of cells to make tissues.

These substitutions reconfigured Harrison's achievement as the making of a general experimental method called tissue culture. Rather than taking an internal process and making it visible externally, in Carrel's system the internal process was rendered permanently external to the animal body. Tissue culture was, as one of the junior fellows working in Carrel's lab put it, the creation of "a new type of body in which to grow a cell"—it involved the development of an artificial, technological, transparent body that would take over the functions of the obfuscating animal body that had been cut away.[5] At first, tissues in culture were clearly related to particular bodies, or at least to the bodies of particular disciplines.

In Harrison's work, it was the embryonic body of development and differentiation; for Carrel it was what we could call the "surgical body," of wound healing and transplantation. However, tissue culture took on a "life of its own" as an artificial body with different characteristics from the animal body.

One characteristic that set tissue cultures apart was immortality, but immortality is not a thing that can be achieved immediately (how could you know a being is immortal right away, without waiting?). The fundamental separation of the body of the organism from the excised tissue culture was achieved most powerfully by the effort to make life *in vitro* "permanent," a quality that cells did not possess in the body. By the fall of 1911, one year after beginning experiments with tissue culture, Carrel was able to keep tissues alive and growing for anywhere from three to fifteen days. He was not satisfied with this achievement. He had already shown that a tissue could be, as he put it, "reactivated" by taking it out of its initial plasmatic medium, washing it in physiological solution, and putting it in fresh plasma. Now he changed the term from "reactivation" to "rejuvenation" in describing the promise of subculturing:

It may easily be supposed that senility and death of tissues are not a necessary phenomenon and that they result merely from accidental causes, such as accumulation of catabolic substances and exhaustion of the medium. The suppression, then, of these causes should bring about the rejuvenation of the arrested culture and thus increase considerably the duration of its life . . . The rejuvenation consists in removing from the culture substances that inhibit growth and in giving to the tissues a new medium of development.[6]

It is perhaps not surprising that a transplant surgeon, well versed in the importance of the maintenance of circulation for anabolism and catabolism, the provision of nutrients and oxygen, and the carrying away of waste products, would see excised tissues as also needing this kind of continual refreshment. What is interesting here is the language of arrest and duration—the culture is arrested by the accumulation of waste products of metabolism and released from this state by the provision of fresh medium. The utilitarian nature of such manipulation suggests that senility and death of tissues are not a necessary fate but an accidental one; duration of the culture's life was the means to test for necessity.

This technique of "rejuvenation" involved taking coagulated plasma that contained the original fragment of tissue and the surrounding new cells, washing it in sterile saline solution, and then placing the fragment in new plasma. The moment chosen for this transfer was the time at which the rate of growth decreased or "large granulations appeared in the cytoplasm of the cells," indicating the onset of "senility." With these experiments, Carrel extended the life of embryonic chick tissues to thirty-seven days. He had thus moved from reactivation to rejuvenation, calling death a "contingent phenomenon"; by 1912 he had proceeded to "regeneration" and then to "permanent life" in describing an increasingly elaborate procedure for maintaining life for increasingly longer periods of time *in vitro*.

Carrel attempted to set up a system of "artificial circulation" in which tissue fragments were grown on cellulose through which a slow stream of serum moved, but this proved to be too complicated. It was easier to move the cultures from medium to medium than to establish a circulation system for the cultures. This he did by growing the tissue fragments in plasma on tiny pieces of silk veil. When the old culture was placed in new medium, the cells would wander out from the old plasma into the new plasma, and the center would eventually die. The results were far from precise; the size of the cultures was erratic and influenced by many mechanical disturbances such as the folding of the plasma clot during handling, and it was difficult to tell whether the mass or the density of the culture was changing. However, the main point of the published reports of this work was again a sense of possibility: permanent life outside the body. This was illustrated particularly well in cultures of embryonic heart cells, in which the resuscitation of the cultures could be observed in the ability of the fragments of heart muscle to beat.

Cultivation of the Heart (Experiment 720–1) On January 17, 1912, a small fragment of the heart of an eighteen-day-old chick fetus was cultivated in hypotonic plasma. The fragment pulsated regularly for a few days and grew extensively. After the first washing and passage on January 24 the culture grew again very extensively, but there were no rhythmical contractions. On January 29 and February 1, 3, 6, 9, 12, 15, 17, 20, 24, and 28, the culture underwent eleven washings and passages. It became surrounded by fusiform cells and many dead cells. There were no pulsations. After the twelfth passage the culture did not grow at all. Then the tissue was dissected and the old plasma was completely extirpated. A small central fragment was removed, washed and put in a new medium. On March 1 it was pulsating at a rate that varied between 60 and 84 per minute. On March 2 the pulsations were 104 at 41° C., and on March 3 80 at 40° C., but on March 4 the pulsations were very weak and stopped altogether at 2 P.M. On March 5 the culture underwent its fourteenth passage, and the pulsations reappeared immediately.[7]

This excerpt of the results section of "On the Permanent Life of Tissues Outside of the Organism" offers the heartbeat as proof of the power of regeneration that Carrel was claiming for his method of serial cultivation. Although he cultivated all kinds of embryonic tissues, he chose heart tissue to illustrate the possibility of endlessly renewed life, with its highly manifest liveliness—the rather uncanny ability to pulse, stop pulsing, and start again over the space of several days. An undertone of resurrection pervades the short sentences reporting the weakening, cessation, and reappearance of pulsations accompanying the change from old medium to fresh. Although at the time this paper was published, the tissues had lived for only two months, he used their continued life as evidence of the possibility that "the length of the life of a tissue outside of the organism could exceed greatly its normal duration in the body, because elemental death might be postponed indefinitely by a proper artificial nutrition."[8]

There is evidence that many who tried their hand at tissue culture also chose to begin with heart tissue. The sight through the microscope of the pulsating tissue was apparently singularly affecting. Henry Field Smyth reported in *JAMA* in 1914 that he had attained "most satisfactory growths from seven-day to eleven-day hearts," which "pulsate so violently that they are apt to tear loose from their plasma supports."[9] The cytologist William Seifriz wrote to Albert Ebeling, Carrel's longtime assistant, in 1933, noting that he had over the years used heart tissue to teach the techniques to students, because "the continued beating of a heart fragment is one of the things that the student always enthuses over." In fact, he wrote, "I must admit the same naïve feeling myself every time I see it." Seeming defensive that Ebeling would think him "altogether too childish," he hastened to add that he thought that other scientists felt the same way: "If I remember correctly [Warren] Lewis was no less enthusiastic when he told me of the pulsations which he had observed in a single cell."[10]

What Carrel failed to communicate in this highly optimistic appraisal of the chances of suppressing death *in vitro* was the fact that, although the cells continued to multiply at each subculture, the cells themselves often got smaller and smaller with each transfer. Thus, although the tissue lived longer and manifested important signs of life—movement and multiplication—it ceased to increase in mass. Carrel let this be known only after he found a methodological solution to both erratic growth patterns and the lack of increase in mass of the cells. He found that the addition of extracts made of ground-up tissues, in particular ground-up embryonic tissue, solved the problem.

Carrel termed these embryonic tissue extracts "embryo juice." Apparently inspired by the thyroid gland pulp that had made skin wounds on dogs heal faster in earlier experiments of 1907 and 1908, he ground up adult and embryonic tissue with sand in a mortar. He added saline solution, put the tubes in cold storage, and centrifuged them. The supernatant solution was then filtered through paper and added to the cultures. He found that embryonic extracts "activated" growth in these cultures much more than extracts of adult tissues, although spleen tissue and cancerous tissue were also growth activators. He took the rate of growth by measuring the ring of new tissue around the original fragment with a micrometer. He found that fragments of tissue, as long as they were approximately the same size and taken from the same tissue of animals of the same age, would grow at the same rate in his new culture medium of two parts plasma and one part embryo juice.

The search to find a nutritive medium that would keep cells alive and growing without diminishing in size was, like other technical changes, also interpreted as an increase in knowledge about the nature of cells in general. By 1913, Carrel was declaring a "constant relation" between "the rate of growth and the composition of the medium," and this discovery opened a new field of investigation—cell-medium interaction—described as follows:

> Certain cell phenomena of the higher animals, such as multiplication, growth, and senility, might now be investigated profitably. Since the time of Claude Bernard it has been known that the life of an organism is the result of the interactions of the cells of which it is composed and of their milieu intérieur. But the nature of the interactions has not yet been ascertained; for in order to discover the laws by which they are regulated it would be necessary to modify the humours of the organism and to study the effect of these modifications on the growth of the tissues. This could not be done on account of the lack of a proper method; but this investigation is now rendered possible because of a technique which permits strains of connective tissue cells to multiply indefinitely *in vitro*, like microorganisms.[11]

By putting fragments of tissue of "known activity" in different media, and putting tissues of different activity in media of the same composition, he wrote that he had found the scientific method appropriate to analyzing the *milieu intérieur*. This was a very explicit connection of the methods of microbiology, with its ease of access and manipulation of its unicellular living subjects, with those of physiology, as classically defined by Claude Bernard: destructive intervention or interruption of life processes to see how they worked, as they worked.

In short, this was a statement of the possibility of taking vivisection outside the intact body. One could modify the activity of cells by purposefully modifying the medium, or one could analyze the medium for the changes induced by the cells living in it. The "known composition" of the medium was rather misleading: "Embryo juice" was hardly a clearly delineated substance of known composition. However, all attempts to use purely synthetic media for culture, such as saline solutions, had resulted in limited life spans for the tissues. It seemed that "embryo juice" was the key to the prolonged, even indefinite, life that Carrel was now claiming for his cultures.

The discovery of workable nutritive medium for the prolonged maintenance of *in vitro* life was thus bound up with claims for tissue culture as a methodology for the external study of interactions between tissues and fluids inside organs and muscles. By September 1913, Carrel had maintained the original embryonic chicken heart culture for sixteen months of "independent life" over 190 passages. It had stopped pulsing after 104 days, but it continued to proliferate at a rate equal to fresh connective tissue taken from an eight-day-old chick embryo. "It appears, therefore, that time has no effect on the tissues isolated from the organism and preserved by means of the technique described above."[12] After the chicken heart culture had lived longer than the life span of the average chicken, the final terminological shift was made from permanent or indefinite life to immortality. As he would write later, after translating his initial results into the language of Henri Bergson's philosophy: "Time is recorded by a cell community only when the metabolic products are allowed to remain around the tissue . . . If these metabolites are removed at short intervals and the composition of the medium is kept constant, the cell colonies remain indefinitely in the same state of activity. They do not record time qualitatively. In fact, they are immortal."[13] Thus the possibility of immortality was introduced into tissue culture. In this context, immortality was never a given; it was always something that had to be attained by technical means and a particular protocol. These technical means in turn had to be directed toward the exploration and control of biological time.

Wound Healing and Biological Time

Much of the interest and enthusiasm about tissue culture were put on hold during World War I. The Nobel Prize committee considered Ross Harrison for the 1917 award in medicine, specifically for his initiation of tissue culture techniques; but due to the war, no prize was awarded. By the time the committee returned to its task, the novelty and some of the promise of tissue culture had dissipated. Carrel stopped working in the laboratory, though he left his assistant Albert Ebeling to tend to cultures and conduct experiments. Carrel was in France when World War I began, and he stayed to offer medical service. With the financial backing of the Rockefeller Institute, he set up a research hospital at Compiègne, Front Hospital No. 21. The work he did at that hospital influenced the way he thought of his tissue cultures and is thus worth a brief mention.

Having toured several hospitals within a few kilometers of the firing line of the French front, Carrel was "very much impressed by the frequent occurrence of gangrene, suppuration, and infections of all kinds." Indignantly, he asked whether "we surgeons have really progressed with a rapidity at all comparable with that of armament makers and engineers," as so many wounds ended in amputation, infection, gas-gangrene, and finally death.[14] Together with Dr. H. D. Dakin, Carrel set out to find a nonirritating antiseptic and a method of its application that would prevent this kind of infection of deep wounds.

Despite their eminently practical application, Carrel's prewar research interests were not forgotten in his work with war wounds and antiseptic solutions. Carrel put the young, mathematically trained Lieutenant Pierre LeComte du Noüy to work measuring the surface area of healing wounds. Carrel, convinced from his prewar work on dogs that the healing of surface wounds maintained under antiseptic conditions proceeded according to a geometric law, had been trying to estimate the surface area of wounds following the incredibly laborious and rather inaccurate method of placing cellophane over a wound, tracing its outline with a wax pencil, transferring the outline onto a sheet of paper, cutting the paper along the outline, and then weighing the paper. Du Noüy repeated the tracing part of the procedure. Then he calculated the area of each tracing with a planimeter and began to make curves relating the area of a wound to the time it took to heal—an activity of patience and a steady hand that seems incongruous given Du Noüy's report that the work was done "rocked day and night by the unceasing bombardment which formed a sonorous background to all our thoughts and all our actions."[15] He was soon able to chart the rate of healing over time as a function of the size of the wound and the age of the patient.

Carrel and Du Noüy found that wounds of the same size healed more rapidly in younger men and in fact that the rate of tissue repair was twice as fast at the age of twenty as it was at the age of forty—which was the age range of their unfortunate subjects. One outcome of this work was an effective method of sterilizing wounds that was gradually adopted into military medical practice. Another was Carrel's turn to experimenting with the body and its tissues as—in the Bergsonian language that he would increasingly adopt—a register in which time was physically inscribed. After the war was over, *in vitro* cultures became the register to hand; the method of wound measuring, he wrote, was unfortunately "not practical because it requires the presence of a wound."[16]

The return from World War I to laboratory work marked a change in Carrel's focus; he did very little surgery and instead concentrated on improving tissue culture methods for a number of years. The original hanging-drop technique had quickly proved unsatisfactory. The tissue growing in the clotted drop of plasma was in a clump, not a single plane; and thus it was difficult to focus on through the microscope. Furthermore, the cultures were limited to a very small size, and it was difficult to change the medium surrounding the cells, as they became embedded in the plasma clot. Bacterial contamination was a constant problem, and each handling of the tissue exposed it to contamination.

After working on it for some time, Carrel introduced a new form of culture vessel of his own design. This vessel, later known as the Carrel flask, was widely used well into the 1950s. It was a small, flat, round flask five or eight centimeters in diameter, with a narrow, oblique neck. The shape of the neck prevented contaminants from the air from falling directly into the flask when it was open, and the neck could be flamed before and after tissue and medium were placed inside. Tissues were grown in a thin coagulated layer of plasma or fibrinogen on the bottom of the flask and bathed in a liquid medium. The medium could be added and removed through the neck of the flask with a pipette or aspirating needle connected to a vacuum apparatus, or the medium could be similarly aerated through a needle, without disturbing the tissue in the solid medium. Special long-handled tools were designed to al-

Figure 3   A tissue culture worker demonstrates the use of a Carrel flask, drawing off the old culture medium with an aspirator. Undated photograph, circa 1923. Special Collections, Lauinger Library, Georgetown University.

low manipulation of the tissues inside the flask through the neck (see Figure 3).

Emphasis was placed on the way the shape of the vessel both ensured asepsis and allowed the work of tending the culture to be performed in a quick and efficient manner. As in surgery, asepsis, speed, and efficiency were all interrelated in determining a successful outcome.

The fluid medium must be changed every 2nd, 3rd, 4th, or 5th day, according to the nature of the medium and the tissues. The flasks are brought into a room where the air has been sprayed and is practically free of dust. The rubber caps are removed and the neck is carefully flamed. Then the fluid is withdrawn by means of the aspirator or a pipette, and the new fluid introduced. The neck is again flamed and closed. The time required for changing a complex medium varies from 45 to 75 seconds. It is generally possible to handle about 60 flasks in one hour.[17]

Finally, the shape of the flask and the materials used in it were selected for optimum optical transparency. Carrel flasks had several minor variations. One had a bottom opening closed by a thin mica plate; the flask could be inverted and directly slotted into a microscope for high-magnification studies or photography of the living cells as they grew. Later the flasks were refined so their flat glass surfaces were thin enough to be used with oil immersion lenses.

Thus the flasks were not just glassware; they were an integral part of a method in which the fluid medium could be regulated and changed at will without perturbing the tissues and the culture could be observed at any time. The drive to improve the method was, Carrel reported, to make possible the "use of pure strains of cells in a known condition of activity, and of media of almost unvarying composition," as opposed to the hanging-drop preparations, in which "the cells are subjected to complex and obscure influences such as those of necrotic cells of their own type, living and dead cells of other types, and a medium which deteriorates spontaneously within a short time."[18] His primary goal in developing the method, besides asepsis, controlled conditions, regulated medium, and constant access to observation in the living state, was a controlled manipulation of time, both in the longevity of the cultures themselves and in the time of experimentation.

A cornerstone of the new tissue culture apparatus was the time-lapse microcinematography used to record the ongoing lives of the cells kept in these transparent flasks. Carrel did not develop microcinematography himself, nor was he the technology's most proficient user; but it was an important element in what he called "the new cytology"—a science that studied cells as dynamic, temporal beings rather than as static, killed entities of histological staining. Cinematography was one tool, along with the glassware and the apparatuses for changing the fluids around the cultures, that helped materialize biological time as a thing that could be physically intervened in as part of experimentation with tissue culture.

It is not a coincidence that a method developed by Harrison to watch living cells as they changed over time would seem highly appropriate subject matter for the new technologies of cinematography being applied to many realms of the physical world, particularly the microscopic realm. In 1912, Carrel went to Paris to see Jean Comandon, a medical researcher-turned-filmmaker who was being supported by the Pathé Brothers film production company to make microcinematographic films. In 1913 Comandon, in collaboration with two other biologists who had recently taken up the techniques of Harrison and Carrel, produced the first film of cultured cells, *Survival of Fragment of the Heart and Spleen of the Chicken Embryo: Cell Division.*[19] This film demonstrated the potential of cinematography for observing very slow, very small movements. Ross Harrison had noted that he had had difficulty capturing the movement of nerve endings in culture by sketching: Although the nerve endings moved constantly, the rate of movement was so slow as to be almost imperceptible with the naked eye. With a camera, an image could be taken through the microscope at regular intervals of seconds or minutes. When the film was then projected at sixteen or twenty-four frames a minute, the very slow movements were greatly accelerated, making previously imperceptible behavior clearly visible. These movements were also greatly magnified by their projection on a movie screen, the films could be shown to

CULTURING LIFE

large numbers of people at once, and a single movement or a single film could be viewed repeatedly or backwards. This new flexibility of viewing time and the temporal manipulability of the film itself lent to the sense of control over biological time granted to the experimenter by the apparatus of tissue culture.

Carrel did not install his own film-microscope setup until after World War I, but it became central to his later work on "the new cytology." Writing in *Science* in 1931, in characteristic style he dismissed the previous century of cytological study as an incomplete science due to its concentration on the study of form, to the expense of the study of function. The old cytology, Carrel wrote, had "considered cells and tissues as inert forms" because they had been excised and viewed in a fixed and stained state. He compared this to the dissection of corpses because rendering cells visible in this way also meant killing them: "Dead organs and histological sections are nothing but useful abstractions. The body really consists of a flux of structural and functional processes, that is, of an uninterrupted modification of tissues, humors and consciousness."[20] The practices of classical histology had only served to strip cells from their reality by abstracting them from both space and time. The old cytology had placed the emphasis on the static building block nature of cells and tissues, whereas the new cytology studied them as active agents, as "the *builders* of an organism capable of developing, maturing, growing old, repairing wounds and resisting or succumbing to diseases."[21]

To remedy the faults of the old cytology, Carrel wrote, "one must return to the close observation of the concrete event which a tissue is." The cornerstone of the new cytology was cinematography: What other method could capture an event unfolding over time? "A tissue is evidently an enduring thing. Its functional and structural conditions become modified from moment to moment.

Time is really the fourth dimension of living organisms. It enters as a part into the constitution of a tissue. Cell colonies, or organs, are events which progressively unfold themselves. They must be studied like history."[22] Cinematography was the only medium capable of capturing the dimension of time, but the surrounding apparatus allowed the observation of cells without the need to remove them from their *in vitro* existence. The thin-walled flasks simultaneously housed and exposed the cells.

Carrel used cell cinematography as the foundation for a theorization of biological time as completely distinct from clock or solar time. What tissue culture on film provided was direct access to the constancy of change over time; to Carrel, the films revealed that time was actually part of the very constitution of cells and tissues. What one witnessed on film, he said, was "physiological duration." Carrel meant "duration" as he interpreted Henri Bergson to mean duration, attending in particular to the emphasis on ceaseless change with concurrent accumulation in Bergson's writings: "the present of a living organism does not pass into nothingness. It never ceases to be, because it remains in the memory and is entered in the tissues. Bergson has clearly shown how the past persists in the present. The body is obviously made up of the past."[23] With cinematography, Carrel thought he could see duration; and if it was possible to see duration, it must be possible, he thought, to measure it, to read its speed by charting the inscription of time into tissues, to find the mechanism by which the body was made of the past—and then to manipulate duration to the point of suspending it altogether. This was perhaps the most hands-on interpretation Bergson's work has ever received; it was a science of duration complete with its own glassware, instrumentation, choreography, outfits, lighting, and atmosphere (see Figure 4).

Certainly someone with Carrel's intimate knowledge of the cir-

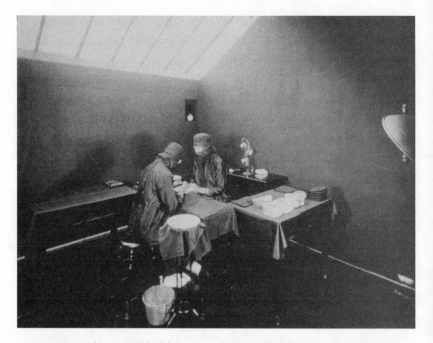

Figure 4 Alexis Carrel's laboratory at the Rockefeller Institute for Medical Research was lit from above with natural light. The walls were painted gray and technicians and scientists dressed in full-length black gowns to increase the visibility of the tiny, translucent pieces of tissue they worked with. Undated photograph. Special Collections, Lauinger Library, Georgetown University.

culatory system, accumulated over years of stopping, starting, and redirecting the flow of blood through the body, would not need cinematography or a philosophical understanding of duration to reveal the importance of fluid replacement for life processes. However, he interpreted the details of movement and behavior captured on film as visual evidence of the constant interaction of the cells and fluids of the body as well as the mechanism by which time was literally inscribed in physiological matter. Continuing the work he had done during the war on the relation between age and wound healing, Carrel bathed his constantly dividing chicken heart culture in the blood serum of animals of different ages. Although the serum of a one-day-old dog did not affect the rate of cell division, Carrel discovered that the older the animal was that provided the serum, the slower the rate of growth. This too was evidence of the body as an open register in which time was constantly being inscribed.

Carrel saw cells *in vitro* as a simple animal representing the closed system of tissues and organs bathed in blood and interstitial fluid.[24] The flasks he designed were the boundaries of these simple, closed systems; the experimenter controlled what went into and out of this artificially isolated portion of space. Tissue culture made physiological duration visible: "Physiological duration . . . appears as soon as a portion of space containing metabolizing things becomes relatively isolated from the surrounding world."[25] For these cells, physiological duration was composed of metabolic processes that created products that changed the cellular medium. The buildup of metabolic byproducts equaled the buildup of duration. From this assertion it was a very short step to the alleviation of time. "If . . . composition of the medium is kept constant, the cell colonies remain indefinitely in a same state of activity. They do not record time qualitatively. In fact, they are immortal."[26] In other words, to put it in the prosaic terms of laboratory equipment, the regular aspiration of the medium and the addition of new medium ensured continued life, because there was no physical registration of time within the isolated space that was the interior of a Carrel flask.

Immortality was for Carrel a scientific concept, and scientific concepts are, he wrote, "operational concepts; in other words, concepts equivalent to the set of operations by which they are acquired. And those operations depend necessarily upon techniques."[27] Carrel borrowed and often quoted this understanding of the scientific concept as that which must "involve as much as, and nothing more than, the set of operations by which it is determined," from the physicist P. W. Bridgman's 1927 *The Logic of Modern Physics*.[28] Following this definition, the scientific concept of immortality was equivalent to the set of techniques that made up tissue culture. Carrel, in approaching immortality as an "operational concept," not only sought to interpret his results in terms of Bergson's concepts of time and duration but also developed a set of instruments and practices as an explicit materialized equivalent of these concepts in the form of a "set of operations."

Although this understanding of *in vitro* life as a philosophical problem as well as a biological one earned Carrel a reputation for eccentricity and mysticism, it was always tied to material and technical results. He saw tissue culture as the "experimental field" in which questions of organization, differentiation, and regeneration could be addressed by observing the behavior of cells. He carried out a long series of experiments attempting to characterize different cell types by their mode of locomotion, behavior toward other kinds of cells, level of activity, life span, and nutritional requirements, on the presumption that "each type of tissue appears to record time in its own way"—which was a recognition of the body as composed of cells with different speeds, life spans, and metabolisms, with a heterochronous existence for different parts of the body.[29]

However, the most enduring proof of his claim to have understood how biological time works remained the famous "immortal chicken heart." Continuing to grow and divide for years on end, these cultured cells divided endlessly at the same rate, in apparent confirmation of Carrel's assertion that he had removed them from time, or removed time from them. After experiments in the 1960s showed the life span of normal cells in culture to be limited, in contrast to that of cancerous or "transformed" cells, which were then referred to as "immortalized," many writers have looked back and disputed that Carrel's culture of immortal chicken heart tissue could have been what he said it was. Hypotheses have been offered that the culture was continuously reseeded (either intentionally or accidentally) with new cells left in the embryo juice used to feed the cultures, or that the culture was at some point transformed, perhaps by contamination with Rous sarcoma virus, which was in use at Peyton Rous's laboratory at the Rockefeller Institute at the same time. Carrel cultured chicken sarcoma cells from the Rous lab alongside his other cell strains, and cross-contamination could easily have occurred. However, the culture itself cannot be tested, as it no longer exists.

The interesting question is not whether the culture was actually immortal in the way Carrel claimed it was, but why, after the initial storm of surprise and disbelief, colleagues accepted the premise of the immortality of somatic cells without question. In 1927, Ross Harrison called the establishment of immortality in culture the "greatest single achievement in the field of tissue culture."[30] Carrel's colleague Jacques Loeb greeted the result with an "of course" response, taking it as evidence that death was contingent and therefore the thing to be explained, not ongoing life.[31] Raymond Pearl, in a 1922 work called *The Biology of Death,* likened somatic cells in culture to amoeba or germ cells, proof that continuous biological lineages existed in continuous streams across the ages.[32]

Why did no one challenge this claim seriously until the 1960s? There are several possible answers. If other scientists had followed Carrel's techniques, they too could have cultured tissues for relatively long periods of time. No one was, after all, equipped to test for forever. Second, the sense of control over biological matter was very appealing, and it was a period in which human control of biology seemed feasible and desirable. Tissue culture was used both as evidence of the human power to manipulate biology and as a demonstration of the dangers of unfettered reproduction.[33] That is, it seemed to simultaneously represent biologists' ability to manipulate life and the potential anarchical powers of proliferation hidden in biological matter, which necessitated control. Scientists were more likely to see the death of their tissue cultures as failures of technique rather than challenges to the idea of life's indefinite bounds.

The Public Life of the Immortal Chicken Heart

The public life of the immortal chicken heart is both a comic and a grotesque episode of early twentieth-century biological modernism. In a beautiful but probably unintentional spelling mistake, the *Indianapolis News* called the story of Carrel's experiments "grewsome," in reflecting that the story had all the "creeping horror of the most morbid narrative of Edgar Allan Poe, with the additional shiver that it is the truth and not the product of a fantastic imagination."[34] Choices made in the laboratory—glass for transparency to observe the ongoing life processes of tissues, heart tissue to have access to obvious and familiar signs of continued life—gave a very specific form to the scientifically reconfigured immortality that attracted wide public attention. Autonomy, plasticity, and immortality, although implanted as it were in the cells themselves as innate qualities unexpectedly uncovered by scientific investigation, remained linked to the realization that these were technologically enabled forms of life. The technical form was part of their novelty and their ability to disturb. Both the public and scientific imagination of these entities was not just of the tissue itself but also of its scientific setting, its artificiality. Practical culture, as Donna Haraway has phrased it, is cultural practice. Scientific, literary, philosophical, and popular responses to the object of the immortal chicken heart are more than representations of this new object; they are specific responses to its material form and the artifice involved in its continuing existence.[35] These were not Frankensteinian narratives of the creation of life *de novo*. They were reactions to the idea of living matter as part of a technological apparatus, a life associated not so much with creation as with engineering and invention.[36] The material constitution of tissue culture as the juxtaposition of living tissues and their glass bodies was not just the material culture of the laboratory but also the material culture of their public life as well. Another way to put this is to pose the question of what happened to immortality as an object after its capture as a scientific term. As a specific quality of cells in tissue culture, immortality was newly represented in narratives of cellular autonomy and behavior—images of tiny fragments of living matter enclosed in glass vessels and viewed in microcinematographic films. After its emergence as a tangible scientific object, how did immortality's cultural salience change?

As with the HeLa narratives that constitute the second cycle of immortality stories produced by tissue culture in the twentieth century (see Chapter 4), it would be all too easy to understand the immortal chicken heart as a public presence composed mostly of facile sensationalism. Immortality is like that—claims regarding it always seem rather extreme. However, more interesting questions of biology and modernism are also present in these materials. Much scholarly work has been directed at the analysis of the early twentieth century as a time of fundamental reconfigurations of ideas of space and time in concert with new time-keeping, transport, and optical technologies as well as the manifestation of these changes in art, literature, and architecture.[37] Historian Philip Pauly has argued that a "modernist experience" can also be seen in American biology, discerned in the early twentieth-century turn toward practices or "strategy and tactics" of artifice.[38] His examples, such as the work of behaviorist J. B. Watson with mice in mazes and Loeb's work on artificial parthenogenesis, underscore the interventionist rather than the observational focus of these new laboratory practices:

> The most exciting "game" was to get organisms to do and be new things . . . Biology had long been the paradigm of a natural science, one in which applications were restricted in number and scope; when serious biologists began to see themselves as designers and inventors of new things, the conceptual and practical significance of "nature" began to evaporate.[39]

Tissue culture offers an extreme example of getting organisms to do and be new things via the scientist's practices of artifice. Because of the public profile of Alexis Carrel and his work, we have access to his laboratory practices, as detailed in the earlier sections of this chapter, and their presence in public discourse. A close reading of these materials reveals that the discussion or representation of the biological thing and its qualities—the tissue, the cell, the organism, and its life or immortality—never traveled without an accompanying discussion of the technical means of its existence. Carrel's status, as Nobel Prize winner and elite scientist at the Rockefeller Institute for Medical Research—itself a distinctively twentieth-century institution bent on modernizing biomedical research—only enhanced the sense of tissue culture immortality as a very "high" technical form of life. These objects were seen simultaneously as new forms of life and as new forms of science as it impinged on modern life.

From his first transplants of dog thighs in Chicago in 1905, to his death in war-stricken France in 1944, the American news media was always interested in the doings of Dr. Alexis Carrel. That interest rose when he joined the Rockefeller Institute for Medical Research, with its already high public profile, in 1906. However, it was not until 1912 that Carrel and tissue culture really began making news. Carrel's claim to have enabled the "permanent life" of isolated chicken heart tissue coincided with his winning the Nobel Prize. On October 11, 1912, the *New York Times* announced "Nobel Prize of $39,000 Is Won by Dr. A. Carrel" and added a slightly smaller subheading: "Awarded Highest Honor in Medical World for Using Dead to Restore the Living."[40] President Taft, in a ceremony of congratulation in New York, called Carrel on stage after comparing him to Harvey, Koch, and Pasteur.[41] This recognition served as a great stamp of authenticity for his research; the media tended to blur together the surgical techniques for which Carrel was awarded the prize and the tissue culture work that he was publishing so prolifically at the very same time that he won the prize. His work with surgery and tissue culture made headlines throughout America and to some extent France and Germany as well. Often a story printed in the *New York Times* or the *Chicago Tribune* would appear with the same text but different headlines in newspapers across America.

Even before the Nobel Prize was announced, the news media had trumpeted Carrel's claims of permanent life. On May 2, 1912, a widely published story described the chicken heart culture pulsing outside the body. "Animal Tissues 'Live' in a Jar" shouted one headline, followed by a slightly smaller headline: "Pieces of Heart Pulsate There and 'Death' is Postponed."[42] The fact that the news editor put the words "live" and "death" in quotation marks signals that these were the concepts that were deranged by the context of laboratory and glass jar. The article consisted primarily of an extended word-for-word transcription of parts of Carrel's article "The Permanent Life of Tissues Outside of the Organism" in the *Journal of Experimental Medicine,* the publishing outlet for much of the work done at the Rockefeller Institute for Medical Research. At that time journalists would pore through the latest issue of the journal looking for news.[43] The extended quote was prefaced by an explanatory blurb: "Dr. Carrel announces he was able to keep pieces of the heart tissue pulsating rhythmically outside the organ from which they were taken for more than two months. The fragments were preserved in suitable media in glass jars."

The very same story appeared in the *Salt Lake City Telegram* on the same day but with a different headline: "Another Step Toward Finding Secret of Life. Heart Tissue Pulsates for Days After Removal From Body." In the *Bridgeport Standard* in Connecticut, the editor used "Experiments with Pieces of Heart Tissue." The *Milwaukee Journal* seemed to envision a laboratory full of pickle jars: "Heart in Pickle Keeps on Beating. Regular Pulsations Continue Two Months After Vital Organ is Removed from the Human Body." This version of the headline ignored the source of the tissue—chick, not human—demonstrating the ease with which this particular substitution was made. The *New York Times* put one headline on the story—"Heart Tissue Beats Long After Death"—and no fewer than three subheadings: "Dr. Carrel Announces Startling Results of His Experiments with Culture. 'Permanent Life' Possible. Fragments of a Chick's Heart Pulsated Rhythmically Two Months After Removal."[44]

This cluster of articles with various headlines emphasizes the beating of the heart tissue. Everyone could understand the image of a heart beating in a jar (the usual home of preservatives), even if they couldn't envision a hollow slide or a hanging drop or a tiny fragment of tissue seen through a microscope. Carrel for the most part did not let journalists into his laboratory, and he rarely agreed to be interviewed. The articles therefore were based on the scientific paper itself. Carrel consciously chose, from all the possible organs and tissues of the body, to demonstrate "permanent life" and rejuvenation by culture with a tissue that would manifest life most obviously: the beating heart. The combination of this natural animate function that every reader could feel thumping away within themselves and the familiar, everyday inanimate object of the glass jar—or even the pickle jar—in the same apparatus resulted in the distinctly uncanny image of life continuing severed from the body and contained in glass. Thus although the pulsations of heart tissue and the use of glass vessels had clear scientific and technical consequences, they also resulted in a particular public perception of what *in vitro* life was.

The inclusion of inanimate materials in the making of this living thing led to its perception as an invention akin to other contemporary technological innovations. This is evident in an extended article in the 1912 Sunday magazine of the *Saint Louis Post-Dispatch*, whose headline—"Surgeon Transplants Various Living Organs From One Animal to Another"—suggests it is about Carrel's transplant work though it includes an extended discussion of the tissue culture experiments.[45] At the forefront of this discussion was an interview with Thomas Edison, to see what he thought of the matter.

> The wizard of electricity, who is a professed agnostic, but who is anxious to learn what happens beyond the grave, saw in the accomplishments of Dr. Carrel a possible answer to this ques-

tion of ages . . . In an interview with the Post-Dispatch re-
porter in Paris last August, while touching upon the discoveries of Dr. Carrel, Mr. Edison said: "If some day the scientists
arrive at a point where the human body, after life is extinct,
can be thus preserved and after an indefinite time, through
the transfusion of life-giving blood or fluid be brought back to
resume its normal functions, who can say that we may not
learn definitely that there is consciousness after death?"

Although Edison did not have anything much to say about Carrel's work except to muse on the possibility of consciousness after
death, "the wizard of electricity" was consulted as if about a fellow
inventor. This article is accompanied by a drawing of a man being
built of riveted sheets of metal, while white-coated, bearded scientists a tenth his size scurry about on scaffolding around the body
and oversee a crane lifting a giant heart into place. The article concluded with a tale of a live man with a dead man's knee: "Dr. Carrel told of one instance in which a section of 'canned' cartilage was
rushed from New York to Chicago for use in a case of knee disease.
The cartilage was shipped by express in a tiny refrigerator, arrived
safely, and was grafted into the knee, and the patient is now walk-
ing about the streets of Chicago with a living section of a dead
man doing excellent service in his propelling mechanism."[46] The
mechanical refrigerator was an intrinsic part of this tale; the cartilage that it kept in suspended animation, a "living section of a dead
man," became part of the "propelling mechanism" of a live man.
Again, mechanism and living tissue were conjoined in an operation
that confused the boundaries of life and death. "Life and Death in
Marvelous New Light," declared the *New York Sun* about both tissue culture and refrigerated surgical material.[47]

Despite the swirl of interest around Carrel's surgical work and
its connection to the winning of the Nobel Prize, it was tissue cul-

ture and the claim for immortality that literally and figuratively outlived all the other forms of his experimental work. The chicken heart culture was established on January 17, 1912, and thereafter the staff of the laboratory and the news media would celebrate the culture's "birthday," year after year. Because the tissue ceased pulsing after 104 days and gradually came to consist purely of fibroblast cells without any heart muscle tissue, it was actually inaccurate to continue calling it "chicken heart." The label persisted, although the emphasis on the manifestation of life as muscular rhythmic movement faded, to be replaced by a new fascination with the culture's endless powers of proliferation. As the years went on, having tissue that grew at a constant rate under the same conditions, which could then be altered nutritionally or physically, was an intrinsic part of Carrel's attempt to make tissue culture into an accurately manipulable system. Thus he took care to highlight the endless repetition of the same, in emphasizing the ability of the cells to continuously grow and divide at the same rate as well as his own ability to slow or increase this rate of growth by adding or subtracting things from the surrounding medium.

The narrative of endless reproductive power built on an already marked interest in the proliferative habits of single-celled organisms such as protozoa. In France, for example, the naturalist and prolific writer of semipopular scientific works, Edmond Perrier, wrote a long article in February 1912 in the *Feuilleton du Temps* about tissue culture called "Le Monde Vivant."[48] Cell populations, he wrote, were just like the infusoria whose power of multiplication was already well known: "so great that if all of their progeny live for the normal duration of their existence, only one of these tiny beings, only tenths of a millimetre long, produces in one month, according to the calculations of the eminent naturalist M. Maupas, a mass of living substance a million times more voluminous than the sun. What hecatomb that supposes and what fragility!" The calculation of volume, were one able to feed and keep all of the cells produced by tissue culture, became an enduring feature of writings about tissue culture by scientists and journalists even to this day. It was (is) as if the smallest of biological subjects, the protozoa or the tissue cell, carried with it the larger presence of its potential progeny—a ghostly intimation of enough flesh or protoplasm to outweigh the sun, cover New York City, or fill the solar system. Even the most staid of biologists participated in expanding upon this imagination of fleshly volume. The embryologist Ross Harrison, more prone to dry comments on the "oversanguine enthusiasm" of some proponents of tissue culture, comments probably directed at Carrel himself, nonetheless lectured in 1927 that "had it been possible to provide nutriment for all the cells, and to allow them all to multiply, they would now greatly outweigh the terrestrial globe."[49] In 1937 P. Lecomte de Noüy envisioned the same set of cells reaching a volume "more than thirteen quatrillion times bigger then the sun."[50]

Later, in the 1920s, the cinematography of cells had no small role in this partly fearful, partly awed, partly comic narrative of endless, autonomous, and frenetic reproduction. The "swarming" of cells over the field of view, and the rapidity of cell division when watched via time-lapse cinematography, was a feature of all commentaries and part of the experience of watching the film. For both public and scientific audiences, the life projected on screen was a spectacle such as they had not seen before. A Dr. Green, Professor of Chemistry at Leeds University, interviewed as he was about to sail out of New York on the Cunard Line, had this to say about his experience of viewing Carrel's films:

It was one of the most amazing things I ever saw, the film of the growth of the tissue was taken during twenty-four hours and must have involved a vast amount of reel. What takes place in the twenty-four hours is reduced in it to a comparatively few minutes . . . Dr. Carrel introduces immortality in a physicall [*sic*] sense. It is there before your eyes, and so long as this tissue is nurtured and irrigated it will live. It cannot die. Its growth is so enormous that it doubles itself every twenty-four hours, and if it had not been pared down each day since the experiment began it would now be a colossal monster overspreading all New York.[51]

There are two things that are vast and colossal here: the amount of reel and the potential size of the culture. The statement that makes the transition between discussion of the cinematic medium's condensing action on time and the discussion of the actual object of observation is: "Dr. Carrel introduces immortality in a *physical* sense." Films of living cells in culture evoked a visceral response; after all, they were in a very literal sense depictions of the autonomous life of the viscera, characterized by endless, boundless growth and proliferation.

In tissue culture, life becomes characterized by movement, pulsation, and proliferation. This quickly becomes excess; without ending, there is only beginning and growth. The film medium was itself lauded for its qualities of easy storage, reproduction, and repetition. You could play a film again, you could play it backwards, and you could show it to hundreds of people. As Dr. Green put it, the films introduced the notion of immortality—endless reproduction—in a "physical sense." Such statements foreground not just the thing seen but simultaneously the novel technological and scientific means of seeing such things.

A newspaper report of a showing of one such film to the 1929 International Physiologists Conference shows that contemporaries

were impressed by what their form implied about the scientists'
work of observing:

> By substituting an automatic motion picture camera for a sci-
> entist's eye at the microscope, and gearing it to take an expo-
> sure a minute, Dr. Alexis Carrell [*sic*] . . . obtained a film
> which reproduced the unremitting observation of the camera
> while the scientist was attending to other researches. Half an
> hour of his time, spent in watching the film when it was proj-
> ected on the screen, showed what used to require days of
> patient observation alone at the end of a microscope. The au-
> tomobile observations of cell behavior made by Dr. Carrel
> through his motion picture camera, were shared directly yes-
> terday with about 500 scientists.[52]

"Automobile" here describes something that moves by means of
mechanism and power within itself. The observation machine works
while the scientist is doing other things, contracting the patient labor of days to half an hour. The observations could then be shared
"directly" and simultaneously with 500 other scientists. Another
journalist commented that, during this display, it "was not even
necessary for Dr. Carrel to be present." The scientist and his work
were separated, his observations reproducible:

> Cells of microscopic size appeared on the screen in dimen-
> sions of feet instead of microns. Their interior changes could
> be followed in detail from the rear of a fifty foot room as
> they grew and reproduced. The continuous record of their
> movements revealed dynamic changes in the tempo of their
> "dance," as it was called, which became convulsive as they
> split . . . The visiting scientists applauded and examined the

phenomena again by having the films run through the projector once more.[53]

How Immortal Cultures End

Immortality, after Alexis Carrel's integration of it into the new science of tissue culture, became a technical descriptor for long-term continuous lines of cultivated cell populations *in vitro*. Such cell lines could thereafter—and are to this day—be referred to as "immortal" or "immortalized," although the connotations of this term continue to shift as more is learned about why some cells can seemingly divide without end and some go through a finite number of divisions before stopping. As an operationalized concept and a form of materialized philosophy, Carrel's "immortality" designated a set of techniques directed at maintaining life over time through the growth of cells in an isolated transparent space, where their physical fluid and gaseous milieu could be completely controlled as a means of making them live differently in time. Gladys Cameron, who worked with Carrel for a number of years before writing a manual on tissue culture, explained that she chose to emphasize the Carrel technique because "[t]his method is the most exacting and comprehensive . . . It has for its main purpose the maintaining of cultures over long periods of time."[54]

Carrel's legacy was this set of rigorously aseptic techniques and the sense of possibility for the total manipulatibility of somatic tissue in space and time. Later practitioners of tissue culture would critique both the claim to have established an immortal culture of chicken cells and the single-minded focus on longevity over all other potentials of the tissue culture system; however, in this era before antibiotics, fume hoods, and plastic labware, Carrel's perhaps overly exacting techniques nonetheless firmly established the possibility of indefinite life for cells excised from the body. Thus the demonstrated plasticity of somatic tissues—their ability to move away from their original tissues across a flask floor, to continue dividing despite removal from the interior of the animal body—was intricately linked to an ability to artificially manipulate their temporality. Freed from the bounds of the individual body, they were also removed from the limits of the individual life span.

That is, indefinite life outside the body was possible through the place and nature of the laboratory. Immortality thus became firmly fixed to cellular life as a technical form of existence produced by "the new cytology" far beyond the boundaries of the laboratory or scientific discourse. Narratives of life in glass spread through the public imagination, embodied by the object of the immortal chicken heart and helped along by pictures and microcinematographic films of frenetic cellular life, narratives that were simultaneously about immortality and cells and about "the new kind of body" built to sustain them technologically. What came into public view was not just The Cell or its immortality but a "technology of living substance"—a form of modern life inextricably bound up with an interventionist, inventive science of biology and its particular forms of artifice.

Carrel was forced against his will to retire from the Rockefeller Institute in 1939. He returned to France, and his longtime assistant Albert Ebeling took the "old strain" to Lederle Laboratories. In 1942, Ebeling temporarily gained a place in the public view by publishing an article in *Scientific American* on the occasion of the strain's thirtieth "birthday": "[The cells] are in their 30th year of independent life in the wholly artificial environment of laboratory glassware."[55] In the course of "Dr. Carrel's Immortal Chicken Heart: Present, Authentic Facts about This Oft-Falsified Scientific 'Celebrity,'" Albert Ebeling wrote that fantastic legends had grown up about the chicken heart culture. However, he wrote, the "authentic" facts were themselves no less extraordinary.

> Although this tissue's history is sufficiently impressive without embellishment, legends, some of them fantastic, have grown up about it. In these tales Dr. Carrel's original tiny fragment of chick embryo heart-tissue has grown into a large, pulsating chicken heart; or pieces have to be "snipped off" from time to time to hold it in bounds; or it is being kept in a glass jar or on a white marble slab, with the added setting of a group of scientists crowded around intently watching and tending it constantly, day and night! Yet, even though the simple facts lack some of the drama of the legends, they are important and no less interesting.[56]

A picture of a disembodied hand holding a Carrel flask with a small piece of tissue in it was accompanied by the caption "This is *it*—the famous culture as it is kept" (see Figure 5). As with the newspaper account that compared Alexis Carrel's tales told to the American Medical Association with "the most morbid narrative of Edgar Allan Poe," there is a blurring of science and fiction in the article. In the newspaper account, reality is as gruesome as fiction "with the additional shiver that it is the truth"; in Ebeling's *Scientific American* article, reality has spawned fantastic legends that still cannot outdo the simple facts of "the most extraordinary career ever enjoyed by a chick or a part of a chick."[57] Legend and fact fed off each other. Alexis Carrel paid a press-clipping service to collect the popular accounts of his work; the archives contain literally thousands of newspaper articles, magazine profiles, and cartoons. There is no doubt that popular and political narratives of "technologies of living substance" circulated back into the laboratory and the writings

Lederle Laboratories photo
This is *it*—the famous culture as it is kept

Figure 5   This illustration in *Scientific American* showed a
culture in a Carrel flask and included the label "This is it—
the famous culture as it is kept." Source: Albert Ebeling, "Dr.
Carrel's Immortal Chicken Heart: Present, Authentic Facts
About This Oft-Falsified Scientific 'Celebrity,'" *Scientific
American* (January 1942): 22–24, p. 22.

of scientists, just as they produced new representations of immortality.

This particular circuit came to an end in the early 1940s. At
Lederle, the cell cultures were used to study the toxicity of drugs
and germicides, but apparently this life of utility was not enough to
justify the continued labor of maintaining them. In 1946, two years
after Carrel's death, the cultures were quietly discarded.

M A S S

R E P R O D U C T I O N

Through the efforts of virologists to conquer polio in the middle of the twentieth century, living human tissue was for the first time drawn into biomedical research on a large scale. In the late 1940s, John Enders picked up tissue culture as a way of growing viruses in the laboratory; of particular interest to a very wide number of medical scientists was his use of cultured human cells to grow polio virus. Some years after his successful use of human cells for growing viruses, Enders addressed the American Association of Immunologists with characteristic understatement: "Indeed I have come to regard cultures of human cells as a fairly satisfactory substitute in certain operations for the living host which, in the case of this species, is often difficult to obtain in sufficient numbers."[1]

Unlike the corpses long used in anatomy and histology, human cells used to grow viruses were bodily tissues that were extracted alive, kept alive for indefinite periods of time outside the body, and not destined either for preservation by fixation or for transplantation back into the body. Because viruses, unlike bacteria, will not grow on lifeless media such as agar, they must be cultivated in living things. Also unlike bacteria, until the advent of electron microscopy, they were invisible, detectable only by the indirect means of the characteristic lesions or symptoms they produced in infected organisms. How to grow something of such indeterminate character was the major challenge of virology.[2]

Learning to produce virus and determine infectivity by using cells reproducing in culture is a story best known for its short-term ending: the production of a killed-virus vaccine against polio in 1954. However, it may also be told with much more long-term consequences in view: This is how living human tissues became a standard material base for biomedical research and how human cells came to be understood as valuable by virtue of their role as productive entities of other biological things. This profound shift in usage and perception of human materials should be understood in the context of the two previous chapters: The reconceptualization of the cell, and the development of methods for manipulating the life span and reproduction of cells in large numbers separate from individual bodies, provided the stage on which a new set of problems of virus production and human infectious disease could be explored.

Although taken for granted today, it is not at all inevitable that human cells should be perceived as factories whose productive and reproductive capacities could be harnessed to make large volumes of cells and biological molecules. In the course of this fundamental shift in the role of living human tissues in biomedical research, and the development of techniques to mass produce human cells as research materials, the living human as research subject was fundamentally reorganized—distributed in space and time in previously unimaginable ways. These new conditions for human bodies arose in the years immediately following World War II.

The establishment of life *in vitro* was not coincident with the advent of *human* life *in vitro*. Culturing living human tissue was first attempted in the earliest days of tissue culture, but human cells were hard to keep alive, much less establish in continuous culture. Although early observers of tissue culture enthusiastically extrapolated results from animals to humans, the actual growth and use of living human tissue outside the body was of little significance before the 1940s. Carrel in 1912 tried to culture human sarcoma tissue he obtained from a fellow surgeon, but it lived for only a few days.[3] The difficulty he, and many who followed, had with growing human tissues *in vitro* meant that the early techniques were mostly established using animal tissues, which were themselves mostly embryonic. The chick embryo was a particularly favored organism.

However, a series of developments in the period from 1946 to 1952 changed this situation utterly. Tissue culture became much more standardized, from the techniques and media to the living cellular materials in use in laboratories. Prior to World War II, individual laboratories made and cultivated their own cultures; there was limited transfer of actual culture material. Although many practitioners visited each other or sent assistants and students to learn techniques, and reprints and letters were exchanged frequently, cultures were not. Concomitant with the shift to standardization and widespread, large-scale exchange of cell culture materials was the widespread use of living human materials for the first time. It is quite possible that these two developments were mutually causative. Use of human materials in highly medically relevant studies such as polio research meant an influx of both research interest and philanthropic support into the field. The National Foundation for Infantile Paralysis, supported by the March of Dimes charity, poured money into both individual research ventures and infrastructure-building activities such as funding researchers to attend summer schools in tissue culture techniques. In combination with strong government support, particularly in the area of cancer research, this meant a huge increase in funding and infrastructure building for research activities across disciplines in the postwar era.

Accordingly, the story becomes harder to tell, diffused across larger numbers of people, publications, and types of activities. The focus here is on the intersection of tissue culture and virology in the 1940s and the resulting development of human tissue as a material base for research and the production of therapeutic materials such as vaccines. The reconfiguration of the role of human tissue in biology and medicine was a shift to understanding of human tissue as a potentially productive technology.

Cells as Virus-Producing Factories

In the first half of the twentieth century, viruses were kept in laboratories by passaging them through animals or eggs. An animal would be injected with some infected material and in turn would provide infected material to inject into another animal before it died. This meant continuous maintenance of sick animals. Fertilized chicken eggs provided another means: Once injected, the eggs could be kept in an incubator for some time, as the virus multiplied in the amnion. However, as media, whole animals or even eggs are very complex and expensive, and not all viruses that scientists wished to study would grow in all animals or in eggs. The idea of growing viruses in cells in culture was proposed almost immediately after the culture of tissues itself was proved feasible. In France at the Pasteur Institute, Constantin Levatidi in 1913 immediately seized on the method as useful for growing viruses of polio and rabies: "Given that, in the conditions realized by the method of Burrows-Harrison, modified by Carrel, certain cellular elements live and even multiply *in vitro* . . . it occurred to me to use this method for cultivating certain viruses which do not multiply in the usual milieux."[4] In 1928, two biologists in Manchester published a method for cultivating viruses in tissue that involved mincing live tissue and suspending it in serum fluids in a flask. This method, which was significantly simpler technically than Carrel's methods[5] and thus appealing to virologists, came to be called "the Maitland method." Others tried the hanging-drop method, infecting small fragments of tissue kept on microscope slides.

When virologists began to seriously consider tissue culture methods in the 1940s as a viable alternative to the expense, frustration, and complexity of culturing viruses in whole animals or embryonated hen eggs, the idea was no longer new. In the late 1930s, the first human vaccine for yellow fever was produced after several years of work showing that yellow fever virus grown in tissue culture became attenuated and could thus be safely used as a vaccine.[6] However, it was not clear that tissue culture would work for other viruses and the mechanism of growth was unknown; although it was possible to prove that virus survived for a while in culture, and perhaps even multiplied there, yields of virus from most culture preparations were low. It was not clear at that point what the nature of the association between cell and virus was, with some authors even claiming to be able to cultivate viruses in cell-free media and others claiming that no virus could survive without the presence of living matter. No shortage of researchers applied themselves to the problem, including several at the Rockefeller Institute for Medical Research, but in 1940 how viruses multiplied was basically unknown.[7]

A few isolated successes, such as the use of tissue cultures as a medium of passage to create an avirulent form of yellow fever for a vaccine, and the successful maintenance of vaccinia virus in a simple preparation of minced tissue and serum, indicated the potential of the principle of growing virus while growing explanted tissue. However, when John Enders surveyed the literature on viruses and tissue culture in 1940, the three possible methods then in use—hanging drop, Carrel flask cultures, and Maitland cultures—he found that the maximum length of time virus and tissue could be maintained outside the body was eighteen days. Furthermore, there was a pronounced problem with yield, particularly for tasks that required large amounts of virus, such as vaccine production. In 1940, John Enders embarked on an in-depth investigation of the new method of tissue culture published by George Gey, directed at the prolonged cultivation of cells *in vitro* using test tubes that continuously rotated. Here lies the intersection of the kind of work detailed in the last chapter, the intense effort to control cellular life in time, and the new set of problems that would be superimposed on these efforts, the production of viruses outside the body.

George Gey went to Johns Hopkins in 1929 as director of the Tissue Culture Laboratory in the Department of Surgery; his wife Margaret ran the laboratory and shared in most of the work, if not its publication. The Geys' focus was on the use of tissue culture to study cancer, and they were particularly interested in growing human cells and in the challenge of maintaining them alive outside the body for more than a few days. In 1933, Gey published the results of several years of tinkering with an apparatus that would provide a dynamic environment for cultured cells. This was called the "roller tube" technique of cell culture.[8] The cells were grown in test tubes that were rotated very slowly, at about one hour per turn, in a specially built gas-tight cylinder.

It is possible to bring about a sort of washing action on the growing tissue cells by revolving the tube slowly at constant speed, thus allowing the supernatant fluid to bathe them con-

stantly . . . The revolving action of the tube not only permits
the supernatant or washing fluid to come into contact with
the growing tissue cells at definitely controlled intervals, but
at the same time allows a definite period of exposure to any
gaseous mixture desired during the gas phase of the turn,
when the cells are covered by a minimum amount of fluid.[9]

As Gey explained it, this setup was designed to remedy shortcomings of both the hanging-drop and Carrel flask methods in trying to grow malignant cells. In a flask or a hanging drop, the cells
would actually begin to liquefy the plasma clot around them and
then begin to float free in the liquid surrounding the central mass
of tissue. In subculturing, the basis of Carrel's method, a piece of
the tissue was taken and the floating cells were lost. This made
keeping cancer cells particularly difficult. In culturing a fragment
of malignant tumor, the connective tissue cells in the fragment
usually proliferated in the solid medium of the culture while the
more malignant cells came free in the supernatant produced by liq-
uefaction, where they quickly died. Because researchers studying
cancerous cells *in vitro* were interested in those malignant cells with
the ability to liquefy fibrin, static modes of culture meant a loss of
the very cells that they wanted. For all of Carrel's emphasis on
the technical regime of frequent changes in the medium, the culture simply sat in place between interventions, surrounded by still
liquid.

With Gey's method, by contrast, you could select the malignant,
free cells by collecting them from the liquid fraction and culturing
them in a constantly moving roller tube: "no stagnation of the medium occurs, as there is constantly supplied a fluid medium which
rapidly dilutes any toxic products. In the flask method Carrel has
demonstrated the need for a supernatant fluid but no provision is

made for the constant washing action and the re-implantation effect which are so advantageous for permanent strains of tissue, and which are obtained by the roller tube method."[10] This was not only a much more effective method of keeping cancer cells alive indefinitely but also much less laborious because large numbers of cells could be cultured in each tube. Gey wrote that one tube could contain as much tissue as 50 to 100 microscope slides, and thus, "the comparative labor involved in the two methods in order to maintain the same amount of tissue is readily appreciated."[11] The method allowed for the maintenance of permanent strains of many types of cells "in laboratories where sufficient technical help is not available to maintain the large number of cultures necessary when other methods of tissue culture now in use are employed."[12] This statement may have been a further dig at Carrel, who commanded more abundant technical help and infrastructural support than most of his scientific colleagues.

As with Carrel's flasks, the glass containers were designed as vessels to contain living cells and as transparent bodies for the life going on inside them. The roller tubes were either round or hexagonal tubes, whose flat sides made microscopic examination of the contents easier. The apparatus included a gas-tight chamber in which the gaseous mixture could be changed to accurately control the pH of the culture medium. "The equilibrating carbon dioxide mixture allows the use of unmodified body fluids or an almost exact duplicate of their inorganic composition." In short, the method allowed a better simulation of bodily conditions: "The washing action, in a way, simulates the flow of the body fluids." With this technique, Gey was able to culture human cancer cells much more successfully than any of his colleagues, and it is this method that he would later use to establish the HeLa cell line in 1951, which I will address in more detail later in this chapter.

Gey was during this period working in the laboratory of Warren and Margaret Lewis, at the Carnegie Institute, Department of Embryology.[13] The Lewises did a great deal of microcinematography, and they had just elaborated a new concept called pinocytosis with the aid of time-lapse observation of cultured cells.[14] Pinocytosis, a word coined by Warren Lewis to indicate "drinking by cells," was meant to complement the earlier phagocytosis, or "eating by cells." Phagocytosis was coined by Metchnikoff in the 1880s as a term to describe the cellular intake or engulfment of solid particles of matter. Pinocytosis, by contrast, was first seen, described, and coined as a term by watching films of macrophages moving across the cover glass of hanging-drop preparations, away from explants of rat tissue.

The wavy, sheet-like processes or thin membranous pseudopodia of the macrophage series—monocytes, macrophages (clasmocytes), epithelioid cells—are often elaborate and project out from the body of the cell as waving sheets. The curious motions which they undergo are much emphasized by motion pictures.[15]

Lewis observed these cells taking in globules of fluid from the surrounding medium, which he presumed contained proteins, water, and various salts. He was terse about the implications of such cell "drinking," saying only that "the importance of this phenomenon in cellular metabolism and the economy of intercellular fluids seems almost self-evident," and he apparently showed the films of pinocytosis around the country for two years prior to publishing the results in written form. Others, however, were more overtly enthusiastic. An editorial in the *Journal of the American Medical Association* about the films observed that: "Physiologists will find in the

Lewis phenomenon suggestion of a new mechanism of cellular nutrition and mechanical filtration or purification of body fluids."[16]

Lewis wrote that the cells sometimes took in a "relatively enormous" volume; sometimes one vacuole after another, and often
several at a time, could be seen passing from the periphery to the
central part of the cell. Although it was not exactly clear what happened to these vacuoles—they could be seen decreasing in size and
gradually disappearing—they seemed to move in a very purposeful
way toward the center of the cell. He calculated that the total volume of such vacuoles might in the course of an hour amount to
one-third the volume of the cell and in twenty-four hours, several
times the volume of the cell itself. Because the cells did not actually
increase in size, it had to be assumed that the fluid passed out as
well as into the cell.

Once Lewis had noted this phenomenon in macrophages on
film, he also saw it in cultures of rat sarcoma cells and rat carcinoma cells, leading him to think that "pinocytosis may be a much
more universal process than we at present suspect."[17] Cinematography, through the exaggeration of scale and speed, highlighted a
phenomenon that Lewis then began to see everywhere in his tissue
cultures:

Motion pictures of these cells migrating on the coverglass revealed for the first time pinocytosis. Some one would no
doubt have ultimately observed it by following cells in the ordinary manner, for, after we had seen it in the motion pictures, we were able to follow it under the microscope.[18]

This mechanism of interaction between the cell and its surrounding medium revealed an "economy of intercellular fluids," which
pointed to a remarkably dynamic interchange between the inside

and the outside the cell, and a very active role for the bounding membrane of the cell itself.

With Gey's roller tubes, allowance for the dynamism of cellular life depicted by microcinematography was built in to the apparatus keeping the cells alive. The apparatus itself began to move. The emphasis on living subjects over time was part of the realization that the bodily functions of nutrition, gas exchange, pH regulation, and excretion could only be simulated properly if the movement and flow inside living organisms were also mimicked. Carrel's technique of a solid medium covered by a changeable fluid supernatant had provided a simple closed cell-fluid system; the movement all came from the scientist, who sucked out old medium with a pipette and added fresh medium at regular intervals. Gey's apparatus put this system into constant motion. At the same time, it further automated the labor of maintaining the life of cells; instead of changing the supernatant, constant change was provided by the slow rolling mechanism, in vessels large enough to carry out "massive tissue culture."

Where Gey was concentrating on growing cancerous cells, and on keeping those cells in indefinitely proliferating, stable populations, John Enders was interested in the relationship between continued cell proliferation and virus multiplication. He was continuing the trend of investigations carried out by his recently deceased mentor and friend, virologist Hans Zinsser, on the conditions existing in tissue culture-virus systems. Zinsser observed that although the Maitland technique published in 1928 "may be regarded as having provided one of the most important techniques now available for the study of ultramicroscopic virus agents outside the animal body," the principles underlying this practice were completely unknown and the physiological changes taking place in the culture over time were uncharacterized.[19] Various investigators had managed to apply the method to the apparently different demands of different viruses only by empirical tinkering (amount of tissue used, closing the flasks with cotton or tight rubber stoppers), without any understanding at a theoretical level concerning what was actually going on.

By 1940, when Enders took up the issue, eighteen days was the maximum amount of time that anyone had managed to keep virus going in tissue culture. Even though Alexis Carrel had noted in 1928 that tissue culture provided an excellent means of "fabrication" of large quantities of virus and therefore was a potential means of manufacture of vaccine, in that "a chick embryo crushed to a fine pulp is capable of producing as much vaccine as a calf," in 1940 the problem stood much as it had for a long time, with tissue culture perceived by virologists as a dubious and difficult prospect.[20] Some still held out the possibility that viruses might be cultivated on artificial media; even Carrel, while insisting that tissue culture was the best method of growing virus, speculated that viruses were some kind of toxin manufactured by cells themselves. Others, examining the so-called inclusion bodies that seemed to appear in the cells of virus-infected animals, thought that viruses were some kind of intracellular parasite. Thus Enders was setting out on a series of experiments that were simultaneously fundamental biological questions about the "course of events which follows the infection of cells by a virus" and fundamental practical questions of maintenance and yield—how does one produce the agent on which one wants to experiment, in large enough quantities for indefinite periods of time?

Putting forward the fairly simple hypothesis that the amount of virus decreased over time in hanging-drop or Maitland cultures because their constituent cells died, he set out to compare Gey's roller tube method to the older techniques. The experimental question was very basic: Would there be more virus production if the cells were kept alive longer? It was not clear what virus titer and the state of the cells had to do with one another. Did the virus kill the cells, or did the cells die because of the culture method was faulty and didn't supply the cells with what they needed to survive? If the cells were dying because of the method, then their death would lead indirectly to the disappearance of virus, but the end result—a dead culture without detectable virus levels—could indicate either mechanism. Gey's proposal of a new method, focused on creating better conditions for prolonged multiplication of cells, provided a perfect opportunity to tease out the different effects on cells of virus infection and culture technology. The roller tube method provided an excellent means to see what happened when measures were taken to prolong cellular life—if the cells live longer, does the virus also last longer in culture?

Enders's results were unequivocal. At the point of publication in 1940, his cultures of embryonic chicken cells infected with vaccinia virus had been continuously proliferating, with consistently high virus titers, for fifty-nine days. "A constantly high level of virus is only maintained when conditions for cell metabolism and multiplication are continuously afforded," he wrote with his coauthors.[21] These conditions, he emphasized, were "continuously afforded" by strict adherence to a regimen in which the medium bathing the roller cultures was withdrawn and replaced with new medium daily. This changing of the cellular medium, which would turn out to be a key part of Enders's method of virus cultivation, was derived in part from Gey's emphasis on preventing the stagnation of the cell medium and in part from work done in Carrel's laboratory at the Rockefeller. There Raymond Parker had shown that a culture could be kept alive in the same flask, without any subculturing, for a year. Because subculturing—taking a small fragment of an old culture and starting a new one with it—was the foundation of continuous culture, it was something of a reversal to leave the same tissue in place, changing only the medium in which the tissue lived.[22] Parker found that, although some of the cells in the flask died, areas of dead tissue were periodically repopulated: "Peripheral areas that had been dead and inactive for months would suddenly be filled with new life, as it were."[23] Living animal tissue cells had never before been maintained for so long without being transferred; and to Enders these results suggested that changing the medium was essential to continued life of cultured tissues.

Previous investigators may have been reluctant to withdraw the medium regularly because of doubt as to where the virus was residing in a culture—in association with cells, but outside them? In the medium itself? What if one threw away the very agent being cultivated by replacing the medium? In Enders's experiments, a new and very useful principle became clear: More cellular reproduction equaled more virus production. Implied in this statement was also a synthesis of the aims, previously separated, of tissue culturists and virologists. Tissue culturists were intent on the problem of keeping cells and tissues alive *in vitro* indefinitely so they could investigate cellular morphology, development, and behavior as well as eminently cellular problems such as cancer. Virologists, who were primarily interested in an as-of-yet-indefinable, invisible agent that produced disease in animals, were not necessarily interested in long-term cultures or, for that matter, in cells. Enders linked the problem of cellular maintenance and reproduction to the aims of virology: The methods of the tissue culturists "appeared to us to offer a means of *observing for relatively extended periods* the course of events which follows the infection of cells by a virus."[24]

Thus, as with Carrel and Harrison before him, it was the appeal of watching things over time that drew Enders to *in vitro* methods. Gey's emphasis on maintaining permanent strains in large amounts suited the needs of virology. However, the implications of Enders's findings were not immediately clear, partly because he had used vaccinia virus in embryonic chick cells. This was as close as one could get to a standard in virology, because vaccinia was a cowpox used as a vaccine against smallpox. Successful growth of vaccinia was easily attained in animals or eggs. After 1940, the work was laid to one side for a number of years, as Enders and his laboratory members were focused on war-related work, the unexpected death of Enders's wife Sarah Frances in 1943, and then his resignation from Harvard to become the director of the Laboratory for Infectious Disease at the Children's Hospital of Boston in 1947. Thomas Weller, who started working with Enders in 1939 as a medical student, was joined in Enders's lab by Dr. Frederick Robbins. Weller, long after the fact, remembered Enders as rarely participating in work at the bench but scrupulously overseeing experimental design and observing the cultures himself.[25] At that point, Enders and Weller began efforts to grow the mumps virus in embryonic chicken tissue culture. Because mumps is a relatively slow-growing virus, the principle of maintaining cellular life long enough to see the course of infection was much more important than with vaccinia. If the cells did not live long enough or did not continue to multiply, it would have been impossible to measure increase in the virus, which would not have had enough time to multiply. Therefore, the time of the virus life cycle created a particular experimental temporality to which the cell culture system had to answer. Although he used the relatively simpler and cruder Maitland technique rather than the roller tube method, Enders nonetheless carefully applied the principles drawn from the 1940 experiments, regularly replacing the medium in which the tissues were kept.

One could easily come away with the impression that everything significant in the history of science was discovered by accident. Careless mixtures, mislabeled bottles, dilutions off by an order of magnitude, things hanging around in the freezer—all of these have produced apocryphal stories of great but accidental discoveries. In 1948, Weller attempted to isolate varicella (chickenpox) virus in a system exactly comparable to the one he and Enders had used for mumps but using cultures of human embryonic skin and muscle. According to his and Enders's accounts, they had never really thought about serious work on poliomyelitis virus, but they happened to have a few vials of polio-infected mouse brain lying around in the freezer. During the course of experiments with varicella, they decided to add polio to "a few unused" human embryonic cultures. The system designed for mumps and varicella turned out to work excellently for polio. By all accounts, including his own, Enders was not very interested in poliomyelitis; and although his associate Frederick Robbins was funded by the National Foundation for Infantile Paralysis, he was content doing basic virological work as his contribution to their efforts to combat polio.

Previous scientists trying to cultivate polio *in vitro* had decided that the virus was "strictly neurotropic"—it would grow only in neurological tissues. It was a startling breakthrough for the entire field of polio research when the Enders lab easily multiplied polio in human embryonic skin and muscle tissues. At the time, the only other available method of cultivation of the disease agent in the laboratory was injection of infected matter into the brains of monkeys or mice, who would sicken and in turn provide infected matter to be injected into the next animal. This was an expensive, uncomfortable, and time-consuming way to keep a virus and virtually unusable for the purposes of diagnosis of individual cases.

I cannot overemphasize the presence that polio had in American culture during these years, with corresponding public, scientific, and philanthropic attention and support for research on the problem.[26] Proof that it was possible to grow polio virus in non-nervous tissues was startling enough that, in stark contrast with his 1940s work on vaccinia, a great number of people began to pay attention to what Enders was doing and how he did it. However, as he himself reflected only a few years later in his Nobel Prize acceptance speech, the ultimate value of these findings lay in the relocation of the event of infection from the experimental animal to the cell.[27] With tissue culture came the ability to see what happened to cells over time in the process of infection as well as the ability to demonstrate virus infection by changes in the cells' appearance. Observation of infected tissue cultures thus displaced the need to see and measure "infection" with the indicator that had been used up until then—an animal sickened by injection of the virus:

> study of these agents from both the biologic and practical points of view would have been greatly limited had phenomena not been discovered which clearly and accurately indicate the occurrence of viral multiplication within the tissue culture system itself. Thus it is probable that if it had remained necessary to inoculate experimental animals in order to demonstrate virus in the culture, the method would have been largely utilized as a convenient means for the preparation of virus suspensions.[28]

Enders embarked on a series of experiments that were nothing less than a systematic stepwise conversion of tissue culture into a tool of virology, which linked the utilitarian practicalities of production of a phenomenon to a new mode of investigation of a biological problem—in this case, the virus-cell relationship. In the period from 1948 through 1954, Enders and his associates showed that polio virus could be cultured in abundance in human tissues of non-nervous origin and that the virus very quickly caused characteristic destructive changes in the appearance of cells in culture. These results opened the way to the visual determination of infectivity of a sample in a matter of days instead of the weeks required when animals were injected. The experimental animal was very explicitly replaced with the human cell culture, not just for diagnosis but also for production of more virus materials. Enders and his colleagues showed that cultures could produce high quantities of virus to be used as antigen in complement-fixation tests for the diagnosis of polio, that different strains could be typed in tissue culture via the neutralizing action of specific antibodies (in other words, the culture could be used as a little immune system microcosm), that new virus strains and indeed new viruses could be isolated in culture, and that viruses highly specific to humans could be kept in the laboratory by using cultures of human cells.

The specificity of Enders's achievement—and its extremely high public profile due to the national furor over polio at mid-century—has masked the more general implications of his work. One of those implications is the induction of the living human cell into the role of standard research tool. It is not that Enders was particularly good at growing human cells though others were not. Rather, he decided to try growing viruses in human tissue at a point when the ability to culture human cells was increasing dramatically due to the introduction of the roller tube technique and the postwar availability of antibiotics. His high-profile success led in turn to a massive scaling-up of work with cultured human tissues. Enders had a ready supply of human tissue in part because of his location at the Children's Hospital. A great number of the living tissues used in his experiments came from therapeutic abortions, miscarriages, hysterectomies, and circumcisions. Reading Enders and Robbins's laboratory notebooks from this period is, from the standpoint of contemporary sensibility about informed consent and patient privacy, a startling experience. Sometimes the name of the patient is listed along with the name of the surgeon who supplied the tissue and sometimes not; often the diagnosis or conditions of acquisition of the tissue are given (for example, spontaneous abortion, removal of fetus because of tuberculosis in mother, menorrhagia, endometriosis, hydrosalpinx).

In an effort to find tissues that were free of nervous elements so he could prove polio was not neurotropic, Enders used foreskins removed in circumcisions. He got the idea of using this particular source of tissue from a paper published in 1948 on the cultivation of virus in human skin that had been grafted onto the membrane surrounding the developing chick in a fertilized hen egg.[29] The researchers reported that they had obtained the human skin used in the study from circumcisions (courtesy of Dr. C. Everett Koop and the Surgical Staff of the Children's Hospital of Philadelphia). The skin would "take" on the surface of the membrane, its cells actively proliferating. Chick blood would begin to circulate in the blood vessels in the skin as the tissues grew together, and the skin could be transferred from one egg to another; through serial culture, the skin would keep living for four weeks or more. Injected with virus, the skin cells would be infected, making the system into a host for viruses that could not be cultivated in other ways. Blank and colleagues were themselves drawing upon work done earlier in the decade by Ernest Goodpasture, who in turn was drawing on work done at the Rockefeller Institute in 1912 in the laboratory of Peyton Rous on the grafting of fragments of animal tumor tissue on the chorioallantoic membrane.[30] "The obvious usefulness," they wrote, "of vascularized, morphologically distinct human skin, actively growing in a sterile host that does not produce antibodies of its own" inspired their attempt to develop "a relatively simple procedure whereby bacteria-free, growing skin was available which could be kept growing by serial transfer from egg to egg."[31]

In short, Enders's work with culturing human tissues can be traced back to a series of efforts to create a living human research subject that was not a human person. The need for living human material for research was particularly acute in virology, a discipline naturally interested in tackling diseases specific to humans. Unlike other investigations in biochemistry or physiology, which could be carried out on tissues in the short-term period of survival after excision from the body, the relatively long time frame of virus cultivation required long-living human material. Many human disease viruses could be cultivated in animals or eggs, but some could not be isolated or propagated except in human tissue. And while human subjects such as institutionalized children or prisoners were used in preliminary vaccine testing and other research during this period, the need for an alternate way of experimenting with and on living human biological matter was pressing.[32] The fact that the practical and intellectual genealogy of these efforts leads directly back to the Rockefeller Institute in 1912 is not a coincidence; the history of pushing the boundaries of the plasticity of tissues through their transplantation and explantation is what led from human skin in hen eggs to human skin in test tubes.

HeLa: The First Human Cell Line

In both virology and cancer research, it was increasingly clear that the course of disease was bound up with the lives of cells. In these biomedical fields, human disease was of particular interest. The desire to culture human cells was not just a question of convenience or the ethics of experimentation on humans but also one of simplification, akin to Ross Harrison's desire to isolate embryonic nerve cells from the complexity of the "bewildering conditions" that prevail in the body of the early embryo. Tissue culture seemed to promise a mode of experimentation with human material not otherwise possible—an ability to watch what happens during various transformations—noncancerous to cancerous, embryo to adult, uninfected to infected cells.

The first widely used human cell line, HeLa, was produced through just such an effort to see transformation from one kind of cell to another over time. While Enders cultured human tissues for weeks and months—long enough to diagnose infection through changes in the cells' appearance or to produce virus—George Gey sought what was by then a more traditional goal for tissue culturists—permanent strains of human cells for experimental use. In the laboratory of George and Margaret Gey, at Johns Hopkins University Hospital, the ongoing attempt to establish cell lines from human tissues intersected with another research program to determine the relationship between two types of cervical cancer. The first was a noninvasive form of cancer involving only the epithelial surface of the cervix. The second was invasive carcinoma involving the deeper basal layers and leading to metastasis. We now know that the former is a precursor of the latter, but that relationship was still unresolved in 1951. George Gey had been recruited to the project to grow cervical cancer cells in the hope that their life under glass would reveal something about their action in the body. It was this context that a young woman named Henrietta Lacks entered when she sought treatment for intermenstrual bleeding at the

Johns Hopkins University Hospital. A biopsy was taken of a lesion found on her cervix, and her doctors diagnosed it as cervical cancer.

Before a biopsy was taken, however, Lacks's physicians sent her to be tested for syphilis—a detail that should be viewed in the context of American medical history. James Jones has written about the characterization of American blacks as a "notoriously syphilis-soaked race" by a white medical establishment and the role of this perception in the founding of the Tuskegee syphilis experiments to track the course of the disease in untreated black males.[33] Unwittingly enrolled in the medical research going on at Johns Hopkins when she sought care, Lacks became part of the cervical cancer research project when a piece of her biopsy material was sent to the Geys' laboratory without her knowledge or permission.

In 1951, when it became clear that HeLa cells were going to continue growing and dividing unperturbed by their artificial environment, the label of "immortality"—in precisely the sense Carrel used the term—was soon applied to them. Researchers realized that this cell line would do what others would not—continue to grow and divide, quickly and copiously (or luxuriantly, to use the more vegetative term favored by practitioners). The cervical cancer program was quickly superseded by the numerous possibilities afforded by a rapidly and apparently endlessly proliferating source of human cells. Gey, who had just visited Enders in 1950 to talk about virus cultivation, immediately attempted to culture poliomyelitis virus in HeLa cells. Working with Jerome Syverton and William Scherer, he showed that polio would grow in the cells and that their status as a "stable strain of human epithelial cells which is readily maintained and perpetuated by cultivation in vitro" meant they had experimental qualities not available from primary explants of monkey or human tissues, which were at the time the main media for cultivating polio.[34]

The cells multiplied rapidly in sheets on glass surfaces, and they could be disassociated from each other and the glass with the digestive enzyme trypsin. This method provided cell suspensions—lots of individual cells floating around in liquid medium—which could in turn be "readily dispensed in aliquots to test tubes to provide large numbers of replicate cultures." These replicate cultures were of potential diagnostic use, because the cells responded to polio infection in twelve to seventy-two hours, showing both characteristic cell destruction and virus multiplication. In noting the significance of the cell line, Syverton and Scherer noted one of its "noteworthy qualities" simply as "its derivation from man."[35] In addition, they noted, "the HeLa strain is resistant to the adverse environmental conditions that occur during shipments by train and plane for long distances."[36]

Through this association with polio, human cells in culture became subjected for the first time to manipulations directed toward their mass production and wide distribution as a semistandardized research tool. Given the high profile of polio, Enders had many interested visitors and correspondents inquiring about his methods. Thus methods developed by Gey and other early tissue culture researchers reached a far wider audience within medicine and other biological research areas than they might otherwise have done.[37] Enders, Robbins, and Weller communicated their findings in brief technical communications on the growth of polio in non-nervous tissues and the cytopathological signs of infection seen therein, but they also attempted to communicate with general practitioners. For example, they wrote a "pot-boiler review on the use of tissue culture in virus research" for the *Journal of the American Sciences,* and Weller wrote a similar report aimed at a wide medical audience in the *New England Journal of Medicine* because "these procedures . . . have not yet come to the attention of the medical profession in general."[38]

Among the scientists who corresponded with Enders was Jonas Salk, a young aspiring virologist in Pittsburgh who in short order set up his own tissue culture laboratory from scratch. Inspired in particular by Enders's report that polio could be detected and diagnosed in days by watching for cytological changes in infected cultures, instead of weeks by watching for symptoms in infected monkeys, Salk decided that tissue culture methods would be the fastest route to producing a vaccine against polio. He was delighted to find that 200 culture tubes could be prepared from the kidneys of a single monkey when he employed the roller tube method. Salk scaled up production of cells to serve as both a growth medium and a test medium for viruses. Historian Jane Smith writes that, in Salk's hands, "Enders' tissue-culture techniques transformed virus production the same way John Deer's plow and Cyrus McCormick's reaper transformed agriculture."[39] Salk's techniques would lead to the rapid production of a vaccine against polio by 1954.[40]

The agricultural metaphor is an interesting one, connecting the industrialization of cultivation and harvesting of food with the industrialization of the life sciences. It was indeed an era of mass production of cells in culture, breaking through the individualized, idiosyncratic materials and methods of the pre-World War II period; but it would be a mistake to ascribe this change either to Enders or Salk by themselves. A shift in infrastructure was occurring as much as a shift in ideas. Scholarly attention to the commoditization of life today has not yet extended to any in-depth understanding of life sciences research as itself a market for biological products, even though the manufacture, purchase, and use of industrially produced instruments, reagents, animals, and cells is a fundamental characteristic of life sciences in the latter half of the twentieth century.[41] This story gives some indication of the interaction of the development of the biological products industry with the first emergence of the mass production of human matter.

In 1940, Enders wrote to one of his associates who had recently left for another job that "it may bring back to you the well known scene if I tell you [Weller] is at the moment engaged in preparing embryonic extract." By 1954 he was placing orders with Microbial Associates of Bethesda, Maryland, for a dozen tubes of HeLa cells, along with infertile egg ultrafiltrate, Hanks solution (a commonly used nutrient medium for cell cultivation), test tube brushers, hyperimmune guinea pig serum, and various other substances that previously he would have had to construct himself in his own laboratory.[42] By 1964, he was placing orders for human embryonic kidney cells and writing to Microbial Associates to ask that they arrange for a shipment of monkey kidney epithelial cell suspension to be shipped by Air Express and to be transported to his laboratory by taxi, immediately upon its arrival at the airport.[43] Getting one's cells by taxi was quite a different scenario from preparing cell cultures from scratch. The scientists were in large part responsible for encouraging commercial companies to enter the business of supplying tissue culture needs.

What Enders and Salk were able to do was supported by a concerted effort, underway in the 1940s, to consolidate and organize the tissue culture community. By 1940, use of the technique was rapidly growing despite its severe and obvious shortcomings. Tissue culture was still an exacting technique, in particular due to the need for strict asepsis; researchers would have to wait until the advent of antibiotics after World War II for this aspect of working with cells and tissues to change. Despite efforts to make a synthetic medium, no cell culture would thrive without the addition of embryo extract, a totally unpredictable mix of unknown components. Likewise, efforts to grow clonal populations of cells from a single cell and thus obtain as pure a culture of one kind of cell as was possible with bacteria met with continual failure, leading some scientists to conclude that cells of higher animals could not proliferate when isolated.[44] It was difficult to quantify the mobile populations of cells because measures of size—for example, of the diameter of the outgrowth around an initial explant—could not differentiate between growth by cell size, cell division, or cell migration. Tissue culture glassware could be ordered from biological supply catalogs, but practices remained highly idiosyncratic from lab to lab, with each laboratory making its own culture media and establishing and growing its own cultures, often with equipment constructed by individual scientists.

The combination of evident promise and time-consuming, frustrating difficulties is exemplified in the person of Keith Porter, a prime mover in the effort to network practitioners of tissue culture together into an organization that would standardize the method and collect and disseminate information and images about cells *in vitro*. Porter arrived at the Rockefeller Institute just after Carrel had retired, and Carrel's assistants and technicians had moved on when the laboratory was closed down. Porter wanted to make electron micrographs of whole cells and to compare micrographs of normal and malignant cells as part of an effort to visualize the causative agent of chicken sarcoma in those cells. The promise of using cultured cells was the same as it had been for decades—long-term stable populations of different strains of normal and cancerous cells that could be isolated and seen in their living state. He could control for artifacts by comparing across light microscopy, microcinematography, various fixatives, and the electron micrographs as well as between normal and malignant cells.[45]

However, he was faced with having to learn the techniques of tissue culture and construct all of the media and equipment he needed before even getting to the stage of making electron micrographs. He wrote to Raymond Parker, Carrel's former assistant and author of the 1938 manual *Methods of Tissue Culture*, that he wished he did not have to spend all his time dealing with the problems of tissue culture. What he really wanted to do was electron microscopy. However, the quality of the images depended on his being able to culture the cells:

> I sometimes wish that I could get tied up with a good tissue culture lab so that so much of my time would not be taken up with tissue culture problems. The electron microscope end of things is much more my meat. If only some of those boys who control the cash had enough intelligence to see the possibilities they would really get a good crowd of specialists together and in a few years shake some information out of the cell . . . The quality of one's electron micrographs is very largely determined by the methods used in preparing the material. This is probably more true of these tissue cells than most other materials.[46]

Porter's concern with these issues is typical of the motivation behind a meeting held in 1946 in Hershey, Pennsylvania, run by Rockefeller Institute cytologist Albert Claude in the name of the National Research Council on Growth. Forty-one scientists were in attendance, including Ross Harrison, who was asked to kick off the event with a historical overview of the development of tissue culture. At this conference, the Tissue Culture Commission was formed with Porter as its chairman. The declared aims of the commission were to pursue the standardization and provision of prepared media, the preparation of a bibliography of work in tissue culture, and more generally, to be a service organization that would lighten "the load of routine common to tissue culture laboratories and by so doing give the investigator more time for productive effort."[47]

The Tissue Culture Association (TCA) was central to the rapid standardization of both materials and practices. The membership in this organization climbed yearly, and this success catalyzed the formation of the European Tissue Culture Association in 1948.[48] A testing and certification laboratory was set up that suggested media preparations and production procedures to commercial suppliers—Difco Laboratories and Microbial Associates—and certified their products. By 1949, laboratories could purchase media such as chick embryo extract, horse serum, and human placental serum instead of having to attain and test it themselves. A summer school for training scientists in tissue culture began in 1948 (and continued until 1982), out of which grew a *Tissue Culture Course Manual*. The first issue was dedicated "[t]o those who believe that the study of cells living *in vitro* can lead to better understanding of the processes of life."[49] A tissue culture bibliography, also initiated at the 1946 meeting (when the number of references was predicted to be 2,000–3,000) came out in 1953 with 23,000 publications listed.[50]

Many of these developments were aided by financial support from the National Foundation for Infantile Paralysis (NFIP), which provided fellowships to scientists wishing to attend the summer schools to learn tissue culture and underwrote many of the research programs involving tissue culture because of its link with virology. Historians of science Daniel Kevles and Gerald Geison, noting the formative role that the goals of funding institutions had on animal virology in this period, write that between 1938 and 1956 the NFIP awarded 322 postdoctoral fellowships in virology and other fields related to polio. In 1953 NFIP was spending more than twenty-five times as much on polio research than the National Institutes of Health, despite NIH's explicit commitment to funding microbiology. An official at NFIP estimated in 1956 that one-third of the virologists under age 45 in the United States had been trained through NFIP fellowships.[51]

In addition to financial support, the administrative and public relations resources of the NFIP were squarely behind the decision to mass produce HeLa cells as part of the national testing process for the Salk vaccine. A polio vaccine that effectively immunized monkeys against polio was announced by Jonas Salk in 1952; but before it could be used on people, the vaccine had to be tested to figure out things like how much was necessary to provide immunity. In a campaign organized and funded by the NFIP, twenty-three laboratories were recruited to do the testing as rapidly as possible. The scale and rapidity of this testing, which was originally done on monkey cells, was threatened by the insufficient supply of rhesus monkeys. Gey and two colleagues had shown in 1953 that polio infected cultures of HeLa cells, and therefore the NFIP suggested that HeLa cells be grown in massive quantities for vaccine testing. The founder and chief administrator of NFIP, Basil O'Connor, happened also to be chairman of the Board of Trustees of Tuskegee Institute in Alabama; and through this connection, a central HeLa production laboratory was set up in 1953 at the Institute.

Technicians recruited from Tuskegee graduates were trained, special large-scale culture apparatus was designed and built, airlines were contacted, and quality control procedures were set in place. Production was scaled up from single test tubes to racks of tubes, which could be batch sterilized and then inoculated with cell populations with an automatic dispenser. The cells were grown in medium consisting in part of human blood serum, and the directors of the project reported:

> We obtained the considerable amount of human serum we needed from weekly blood collections from a list of donors invited to sell their blood for this purpose . . . At the John A. Andrew Hospital of the Tuskegee Institute there was a regular time on Sundays to accommodate blood donors who included students at Tuskegee Institute, Auburn and Alabama State Universities, soldiers stationed at Maxwell and Gunter Air Force Bases, as well as citizens of nearby towns. Since the blood levels of alcohol and nicotine were critical factors, we required all donors to observe a specified period of restraint in smoking and alcohol consumption.[52]

The first test shipment from Montgomery, Alabama, to Minneapolis, Minnesota, took four days. Between September 1953 and June 1955, the production center shipped 600,000 cultures of HeLa cells. The cells survived well unless exposed to extremes of temperature, either high or low.

The scientists directing the project published papers on the technical aspects of growing mass cultures and shipping them over long distances. Populations of HeLa cells were established in laboratories all over the country, with a comprehensiveness, level of standardization, and rapidity that simply wouldn't have happened without the sense of urgency and monetary and public national support behind the polio effort. The cells were grown in the blood serum of people "invited to sell their blood" in support of a national effort to develop a vaccine against polio. It was a national effort directed at mass production of cells.

Mass Reproduction

Here, mass production should also be understood as mass *reproduction*. The product in this case does not come to an end, because it can be, but is not necessarily, consumed by its use. The scientist who receives a tube containing a cell culture is receiving a reagent or an experimental tool and, at the same time, the means of making more. And indeed, once in the laboratories, HeLa was used for all kinds of purposes other than polio research, serving as a model for the establishment of other kinds of animal and human cell lines. In this way, the HeLa cell became a standard cell and a widespread scientific presence.

This mass reproduction of HeLa was made possible only through science as a kind of national venture, with soldiers, students, and donors to the March of Dimes all participating. The relationship of institutes such as the Johns Hopkins Medical School and the Tuskegee Institute to the patients to which they offered care was structured by the efforts of these places to be places of medical research; in the case of the Tuskegee Institute, that relationship was tainted by the infamous use of its impoverished patient pool for experimentation. This attitude was embodied in institutes' physical layout and departmental organization; Gey's office, for example, was next to the operating room. Each biopsy was viewed as potentially both diagnostic indicator and research material. In the very same period stronger regulatory structures of informed consent and strictures on the experimental use of humans were being put into place as a result of the Nuremberg trials after

World War II, the subject of "human experimentation" began to slide away from that unitary, individual, embodied person whom these strictures were designed to protect and fragmented into complicated pieces such as human cells *in vitro*.

Polio was just the beginning of the story for HeLa. The distribution of the cells was extensive: George Gey is fondly remembered for hand-delivering the cultures to scientists: "He would put his glass tubes containing the cells in his shirt pocket, use his body heat to keep them warm, and then fly to another city and hand them to a fellow scientist."[53] With philanthropic and commercial avenues of distribution added to personal routes of exchange between scientists, a great many people had access to HeLa by 1954, and these scientists began to use the cell line to all kinds of ends. The availability of HeLa and the rapid standardization of tissue culture techniques were mutually reinforcing developments; HeLa itself was used to work out techniques that were then applied to other, newly established cell lines, and the growth in numbers of people and kinds of research activities involving cultured cells meant the use of HeLa across many subdisciplines of biological research. Standardization, training, commercially available media, synthetic media, more concisely regulated populations of cells, a focus on cancer, the use of antibiotics, the increasing use of tissue cultures to grow viruses, and the growth of postwar biology more generally all account for the transformation of tissue culture from a specialty of a few laboratories into the province of many, from an end in itself to a useful and increasingly invisible experimental ground. HeLa, although only one cell line, was through its relative ubiquity both a means of standardization and evidence of the standardized technique's success.

The successful extraction of living cells from the body and their technical maintenance in the laboratory exposed the cells to experimentation, but the site of reproduction of such cells also became the techniques and apparatuses of tissue culture. When these techniques and apparatuses were scaled up and industrialized during the campaign to develop a polio vaccine, the control of the means of reproduction of human cells was placed in a standardized form in a great many laboratories at the same time. It was not just cells that arrived in the mail but directions for keeping them alive and growing more as well as instructions for making or buying the necessary media and apparatus. What also arrived was the sense of possibility: a new form of experimental subject for virus research, for cancer studies, and for exploration of the basics of human cellular life.

HELA

At the interface between medical practice and biological research in 1951, when Henrietta Lacks walked into the Johns Hopkins Hospital in Baltimore, something happened that had been happening for years: Human material was used as a morphological and pathological specimen in medical research. However, this was not a fixed-and-stained, dead histological sample but instead a fragment of living tissue. It was brought to a laboratory whose central goal was to mimic the functions of the body to such a degree that human cells could be grown apart from the body and used in its place in experiments.

The new possibilities for human experimentation and the implications of the new standardized tools and techniques of tissue culture for biomedical research were embodied by the HeLa cell line. The distribution to and presence in laboratories all over the world of what had been a single specimen from one person was an utterly new mode of existence for human matter. Previously, researchers might have worked on a specimen representative of a category of pathology and tissue—say, a specimen of human sarcoma or normal human epithelium. Different scientists in different laboratories at different times would have understood themselves to be working on comparable objects; even though they had different individual specimens, they were of the same *kind*. Epithelium was the same kind of tissue, and sarcoma was the same kind of tumor; morphology and pathology were the grounds for comparison across experiments at different times and places. With the advent of continuous cell lines that could (apparently) be infinitely reproduced and widely distributed, a much more literal way for different biologists in different times and places to work on "the same thing" became possible. Highly controlled lineages of fruit flies and inbred mice had provided this stability in the form of "model organisms"; HeLa was a standardization of human material for research on human biology.[1]

The success with HeLa—whose establishment I described in Chapter 3—and the L cell line, a mouse cell culture established in 1948 by researchers at the National Institutes of Health, inspired widespread efforts in the 1950s to establish other human and animal cell lines. Any individual specimen could become everyone's specimen, provided it could, like HeLa, be cultured *in vitro*. However, no one has ever told origin stories about the life and times of the L cell line or its originating mouse, or the CHO line and its originating hamster. HeLa, by contrast, continues to this day to be the subject of endless repetitive accounts of its origin from the body of Henrietta Lacks—accounts that almost inevitably are accompanied by a grainy photograph of the beautiful, unfortunate woman who died eight months after the initial biopsy revealed a malignant tumor on her cervix. The cells live and the woman does not. They somehow stand for her, she for them; otherwise, this pair of circumstances would not present itself as a paradox, much less one that has generated such fascinated attention from 1951 to today. That one party in this relation should be alive and the other dead creates a dramatic tension that continues to generate scientific papers, newspaper and magazine articles, full-length books, and television documentaries. The resolution of the paradox in these narratives is always the same: The woman and the cells are immortal—the woman through the cells' life and the cells through the woman's death. It is a personification of the woman who died that gains immortality, though the woman's death is necessary to elevate the cells from unremarkable life, maintained in the laboratory, to immortal life.

At first blush these stories might seem simply sensational, a gratified shiver at life being stranger than fiction, immortality a convenient, colloquial way for scientists to communicate with their publics via journalists who never pass up a good immortality story (the thing about immortality, it seems, is that you can tell the very same story ten years on, and it still counts as news). Cynicism or a dismissal of these accounts as part of the simplification or inaccuracy that comes with scientific popularization is too easy. Rather, there is a direct continuity from the scientific papers through to the television documentaries. Personifications of the cell line in the image of the person, whether in *Science* or in *Rolling Stone,* function as accounts of science that we tell one another about what has happened to the human as biomedical subject and about the human relationship to human biological matter. These accounts, focused by the specter of an individual human person, are responses to something otherwise not easily comprehended in narrative: infrastructural change in the conditions of possibility for human life.

In this chapter I discuss infrastructural changes in how disembodied human cells in use in laboratories lived in space and time, and then I show how these altered conditions of existence are expressed in the immortality narratives surrounding HeLa and Henrietta Lacks. Essential to the infrastructural change behind the HeLa story were new techniques for freezing and cloning cells.

Contemporaneous with, but not directly involved in, polio or virology, these developments changed how cultured cells proliferated and persisted as scientific and cultural entities. Freezing cells resulted in a suspension of the flux and variation innate to cell populations going through continuous cell division. The flexibility and ease of transport of frozen materials, which could piggyback on large-scale industries around frozen food and cattle breeding (for which frozen semen was important), meant much greater spatial mobility for the newly mass reproduced cells—a mobility that was key to having any one thing exist in many places in identical form. A technique for cloning cultured cells, producing a population descended from a single cell, was developed in 1948. The ability to make homogeneous clonal populations brought cell culture much more closely in line with the manipulations of bacteria so important to molecular biology and opened up the possibility of genetic studies on complex cells. More broadly, the ability of scientists to clone cells sharpened and made more obvious the profound implications of being able to grow human cells in culture: Vegetative, asexual reproduction became a possibility for human matter separated from the body. For the first time, the intact human body was not the only place of large-scale generation of living human cells.

Cell Lines and Cloning

Perhaps the most debated question in tissue culture in the first half of the twentieth century was whether a single living cell could ever be cultured by itself.[2] The tissue fragments used to establish cultures were heterogeneous mixtures of different kinds of cells; even techniques developed by Carrel to make "pure" cultures of one type of cell or another involved the manual separation of groups of cells that looked different from one another into different dishes—not exactly a precise determination of cell type. Scientists fiercely argued whether the repeated failures to get isolated cells to grow and divide were a reflection of faulty technique or a reflection of innate biological impossibility. Though cells clearly manifested some autonomy, many biologists thought that the cells of multicellular animals had evolved to a level of interdependence sufficient to require the company of other cells in order to live. This was a debate about the difference or sameness of somatic cells and single-celled organisms, about individuality, and about the relation of the parts of bodies to whole organisms.

In 1948, absolutely contemporaneous with Enders's work with growing polio in tissue culture, a group of investigators at the National Cancer Institute, Katherine Sanford, Wilton Earle, and Gwendolyn Likely, showed that single cells could be cultured and coaxed to divide, eventually establishing clonal populations of cells.[3] They used the L cell line, which had been established from mouse tissue that year. Along with HeLa, L was one of the first continuous cell lines to be used in multiple laboratories. Their beautiful experiments involved isolating a single cell in a capillary tube filled with nutrient medium in which other cells had been living—so-called conditioned medium (see Figure 6). Working with results indicating that cells greatly modified the biochemical constitution of the medium in which they lived, Sanford, Earle, and Likely asked whether previous failures to grow single cells were due to the inability of one cell to modify the huge volume of fluid around it. They reduced that volume by enclosing single cells in very fine glass capillary tubes; the microphotographs illustrating their work show a single cell surviving, beginning to divide, and finally a whole population of cells spilling out of the end of the tube. With all cells in the culture descended asexually from a single cell, this experiment finally firmly contradicted the hypothesis of the biological impossibility of culturing single cells, and it brought the somatic cells of complex organisms more into line with experimentation on genetically homogenous populations of bacteria in molecular biology. From this experiment, for example, one particular clone—929, descended from a single cell taken out of the L cell cultures—came into widespread use.

The technique was exacting and difficult, with many tries with many tiny glass tubes to get one clonal population. Other investigators seized on the challenge of culturing single cells, partly because the prospect of controlled, homogeneous populations of mammalian cells was so exciting. With such populations, tissue culture would be transformed into a quantitative tool of biological analysis. Spearheaded by a group of biologists who believed "tissue culture needed the kind of quantitative thinking found in the phage field," the push was on to make systems analogous to the plating and counting systems of bacterial genetics but with complex somatic cells.[4] Indeed, in the words of Theodore Puck, who was deeply involved in developing methods to create clonal cell populations, the aim was to see and manipulate the "mammalian cell as microorganism."[5] Puck, whose background was in biophysics and bacteriophage genetics, sought techniques "for precise quantitative analysis of the survival and growth of *individual* cultured cells that had been submitted to various experimental treatments."[6] Puck and his colleague Philip Marcus, at the University of Colorado Medical Center in Denver, focused on the idea that cells would only grow in medium conditioned by other cells; but they rejected the capillary tube method as too difficult and leading only occasionally to "large-scale colony production from single cells."[7] They set out to "conserve diffusible cellular products," even though they did not know what these products were, by placing the single cell to be cultivated on a kind of elevated pedestal in a Petri dish, above

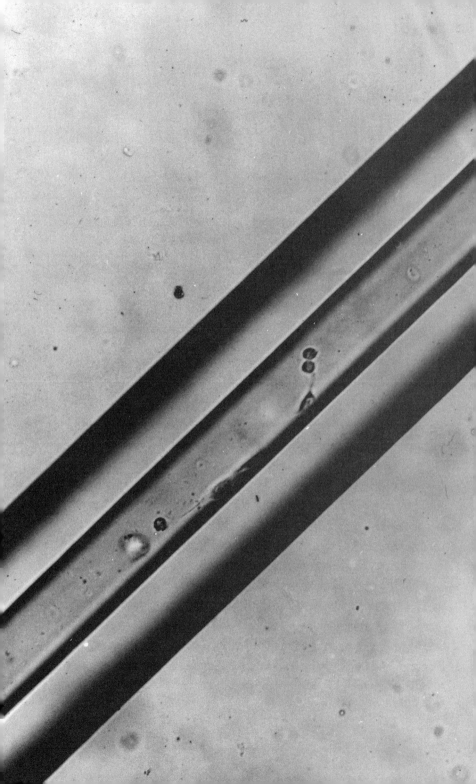

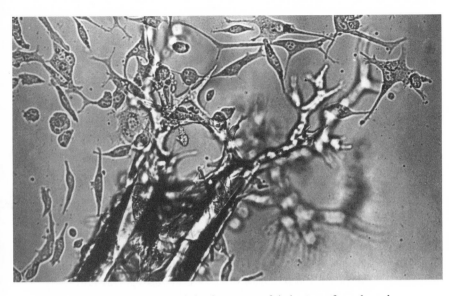

Figure 6   In these images of the first successful cloning of a cultured
mammalian cell taken in 1943, a single living cell rests in a glass capillary
tube surrounded by a tiny amount of nutrient medium (left). In the image above, the population of cells—all descended by asexual mitotic division from the single original—spill out of the end of the capillary tube.
Source: Katherine Sanford, NIH. Photographer: Wilton Earle, NIH.

a layer of irradiated cells covering the floor of the dish. Exposed to
x-rays, these so-called feeder cells were still alive, but they could
not divide; their descendants could not contaminate the colony
formed by division of the original single cell. The feeder cells,
with their continued metabolic activity, conditioned the medium
enough that single cells could be reliably grown into clonal populations; this method is still in use today. The first cells used for this experiment were HeLa cells obtained from the Tuskegee Institute.

    The task of getting isolated cells to grow was intellectually and
practically bound up with the question of cellular autonomy and

interaction. This was prosaically described as the study of the "nutritional needs" of cultured cells; but these issues of what cells consumed and produced spoke to profound questions concerning the relation of cells to each other and to the tissues, organs, body, and metabolic systems of which they were a part. What did cells manufacture for themselves or for each other, and what did they need to have supplied to them from their environment? The only way to disentangle cellular production and consumption was to isolate cells from one another and systematically deprive them of the various components of bodily fluids. Even forty years after Carrel's declaration of "permanent life" achieved by the application of "embryo juice" to cell cultures, biologists did not know what, exactly, made cells grow. They added the body fluids of animals or humans that were processed to various degrees, in various concentrations, but they had never succeeded in synthesizing from scratch an exactly defined medium for growing cells.

In the mid-1950s, Harry Eagle, a scientist trained in biochemical microbiology working at the National Institutes of Health, declared that, by working with HeLa and L cells, he had been able to determine "most of the specific nutrients that are essential for the growth and multiplication of mammalian cells in tissue culture, to produce specific nutritional deficiencies, to study the microscopic lesions thereby produced, and to 'cure' these deficiencies by the restoration of the missing component."[8] Note the physiological language here of the "lesion" produced and cured in the body of the cell by the experimenter. The aim of producing chemically defined medium was not fully attained, but Eagle had taken a big step forward. Immediately, the practical consequences were clear: Use of the "optimum medium as defined for the HeLa cell" allowed Eagle to establish another human carcinoma line directly in medium on glass, without fiddling around with growing the cells first in plasma or fibrinogen clots.[9] This achievement promised to simplify the establishment of cell lines other than HeLa and seemed generally applicable; the medium defined for HeLa also worked for L mouse fibroblasts, suggesting to Eagle that it would be "similarly effective" for a wide range of other human cells.[10]

This kind of work marks the emergence of genetic individuality for the somatic cell—clonal cell lines with distinct, heritable characteristics. The single-cell plating technique of Puck and his colleagues made the determination of the "absolute nutritional requirements of mammalian cells" even more precise. They observed that in nonclonal populations (in other words, those produced from fragments of tissue, not single cells), the ability of cells to "feed" each other masked the differences between "nutritional mutants." Because cell lines such as HeLa and L had been established from a fragment of tissue rather than a single cell and the cells had repeatedly divided and been subcultured, these lines were genetically and morphologically heterogeneous. Cells that required certain externally supplied substances in order to grow—and several such types of cells were isolated from the parental HeLa strain—were analogous to bacterial mutants that could be isolated because of their nutritional particularities.[11] The "standard" tissue culture stocks such as HeLa and L cells (standard although only established a few short years before) were thus shown to be composed of many different "individuals"—once those individuals were isolated, they displayed characteristics different from those of the stock culture as a whole. Not only did they have "marked differences in their nutritional requirements when they are grown as single cells," the populations derived from these single cells kept their distinctive characteristics "throughout continuous cultivation in the same medium for several years and for hundreds of generations."[12]

Although a detailed history of the determination of standardized media is too complex to go into here, what is significant about this research is that it had several important effects on the relationship between researchers and their cells. First, it brought the idea of vegetative asexual reproduction, and the means of that reproduction through single-cell cultivation techniques, within reach of most practitioners. In the short term, this made the cell a candidate for genetic analysis along the lines laid out with research in simpler organisms such as bacteria, which had proved so very productive for understanding the basic mechanisms of DNA replication, expression, and inheritance. In the long term, this was a turning point for the individuality of somatic cells. Long consigned to being part of a group, whether that was a reproducing clump in a dish or a tissue or organ in the body, single somatic cells could became individual entities in their own right, as scientists expanded the particularity of one single somatic cellular genome to a whole population of cells. That individual genetic profile could be singled out for its particular characteristics—its need for inositol or some other metabolically important molecule—and multiplied. Such cells were made into "strains" that were identified and named as entities different from other cells, thereby achieving at least a scientific identity of their own.

Second, the research turned the cell itself into an instrument for the dissection of its own milieu. The capillary tube apparatus was premised on the idea that the cell was not just an entity that consumed things from its environment but also one that produced substances important to its own life processes. These substances seemed both very local and very transient. Puck and his colleagues found that it was not enough to take medium previously conditioned by other cells and expect single cells to grow in it; a microenvironment had to be continuously produced by the feeder layer of irradiated cells. This work led to the identification of many nutrients that were "population-dependent." These nutrients had to be added to the medium around a single cell, but large populations of cells provided these nutrients for each other. These products of cellular metabolism, such as carbon dioxide, certain amino acids, and pyruvate, serve as "substrates for biosynthetic and metabolic reactions that are essential for cellular multiplication, which stops whenever their intracellular levels fall below critical threshold values."[13] Such an idea—that the body was composed of highly specific sets of internal conditions essential to the steady existence of its component tissues and organs—had been present in biological science at least since Claude Bernard's formulation of the importance of the *milieu intérieur;* and, as I described in Chapter 2, Alexis Carrel had put great emphasis on tissue culture being about both cells and the cell-medium system. However, the work of these earlier scientists seemed like rough guesses compared to the precision of using single cells to identify single vital molecules. No biochemical dissection of biological fluids had been as effective as using cells themselves as a tool for sensing and defining tiny amounts of vital substances in their own microenvironments. Things that cells *in vitro* would do or not do (such as divide or survive) could be used to pick specific molecules out of the massive complexity of bodily fluids. Having a homogeneous population of identical descendents from a single cell made this dissection procedure much more precise. Once discovered, the addition or holding back of a particular component could be used to push the cell through a particular process or suspend it metabolically.

Being able to grow clonal populations from single cells in well-defined fluids has, with the advent of human reproductive biology and stem cell culture, attained its own nuances. However, it was indeed "not an accident that the first human IVF pregnancy, Louise Brown in England in 1980, resulted from fertilization of Mrs.

Brown's egg in Ham's F-10."[14] Although this sounds like a particularly British concoction, Richard Ham also worked at the University of Colorado with Theodore Puck on the problem of cultivating individual cells, and he formulated Ham's F-10 medium specifically to grow a Chinese hamster ovary cell line that required pyruvate to undergo cell division. Pyruvate, as it turns out, is a necessary addition to the medium in which mouse zygotes are cultured; otherwise, they will not divide. Thus Ham's F-10 was the medium chosen for human zygotes, which are in the beginning, after all, only single cells in culture.

At the beginning of a lecture published in 1952 called "The Living and Its Milieu," Georges Canguilhem said that "the notion of milieu is in the process of becoming a universal and obligatory means of registering the experience and existence of living things, and one could almost speak of its constitution as a basic category of contemporary thought."[15] This statement appeared just a year before the announcement of the structure of DNA. Soon after, the notion of the milieu became anything but obligatory in explaining living things, particularly in those areas of the life sciences dominated by genetics and molecular biology. Canguilhem's intellectual orientation would soon after be caught up in the stream of code and information, developing into a philosophy of error based on a life science focused on genes and the notion of DNA as language. Suitably enough, however, Canguilhem's thoughts on the series of swings back and forth over the centuries in which the relation between organism and milieu is repeatedly reversed, one determining the other, recurs in a new light today. Now the demands of cultivation of living matter as an integral part of its manipulation again directs attention to cellular media and, more broadly, to the situation of the living thing when the laboratory, biotechnology, and contemporary life science are its milieu.

Cell Lines and Lifetimes

Around the same time that John Enders was fortuitously discovering that polio could grow in non-nervous human embryonic tissue, Christopher Polge and his colleagues in England at the National Institute for Medical Research were accidentally discovering the protective qualities of glycerol for freezing and thawing live chicken sperm. Although it took some after-the-fact chemical analysis to determine what the substance was in the wrongly labeled bottle that had kept the sperm alive through the freezing and thawing process, in 1949 this discovery of glycerol broke open a world of possibility for suspended animation. Sperm were used in many experiments on "suspending animation"[16] because they were easy to get and "expression of their reproductive capacity provides an excellent criterion of functional integrity in survival."[17] In other words, the ability of frozen-and-thawed sperm to fertilize eggs and produce more life was a fabulous measure of whether they had come through the process "alive"—with all their important functions intact.

The use of glycerol was very rapidly extended to other kinds of cells and tissues. One participant in what he called "an early London meeting of bankers and grafters" reported that, by 1953, "we heard a great deal about its preliminary successful application to a variety of cells and tissues. I think all participants left this meeting convinced that, with the aid of this remarkable reagent or other with similar properties, means would be forthcoming in the not-too-distant future that would enable us to freeze and store many different tissues and organs in a state of virtually complete suspended animation indefinitely and very conveniently."[18]

In 1954, William Scherer (who participated in demonstrating that polio virus could be grown in cultured HeLa cells) and

Alicia Hoogasian showed that it was possible to preserve cultured mammalian cells by freezing them with glycerol and storing them at −70°C. They used HeLa and L cells to establish the parameters for freezing and thawing mammalian cells, measuring the growth rate and susceptibility to viruses of the cells upon thawing in order to prove that no significant change had occurred.[19] The cells' morphological appearance and ability to reproduce was "indistinguishable from that observed before storage."[20] As the authors saw it, there were vast practical benefits to freezing cells in terms of facilitating their transport packed with solid CO_2 as well as convenience and protection against loss, change, and contamination.

Freezing also had experimental potential in that it suggested a new mode of comparison. Cells placed in culture seemed to undergo some kind of malignant transformation after continuous cultivation; being able to freeze some of the original culture enabled comparison between normal and transformed cells. Of course normal and transformed cells could be compared with ease, but the temporal rearrangement made possible by freezing suggested a kind of self-comparison of two points in the same lineage not previously possible: the comparison of cells with their own "malignant derivatives." Freezing looped the line in lineage, making two of its points cross for side-by-side comparison. Peter Medawar wrote in 1952 that his principal object in studying the frozen storage of skin was to use it in experiments that would not otherwise be possible, such as the making of an "age chimaera," by grafting tissue from a young animal "to its own self when it has grown older. Such an age chimaera (an organism whose parts are the same genetical constitution but of different developmental ages) can be realized by the appropriate use of storage methods."[21] Freezing thus immediately suggested modes of previously impossible comparison of different points on the same arc of biological time, in the same individual.

Whether this "individual" was an organism or a cell line or strain, one could put the older and the younger selves together in the same experimental moment.

This research into freezing coincided with a flurry of activity in establishing new cell lines and identifying particular strains or mutants within existing cell lines. There was no way to keep cell lines except by continuously feeding them and, when they became crowded or had exhausted their nutrient medium, to subculture them into new dishes (this transfer of a small fraction of the cells to a new vessel is called "passaging"). Scientists quickly noted that serial cultivation led to changes in the cells such as alteration in chromosome numbers as well as changes in behavior or morphology. In addition, laboratories attempting to keep cell lines in serial cultivation were troubled by a number of contamination difficulties; even with antibiotics, cultures could be contaminated by mycoplasma (a hard-to-detect bacterial infection not susceptible to antibiotics because the organism lacks a cell wall) or cross-contaminated by other kinds of cells from other cell lines in the laboratory. Entire cultures could be lost through contamination or accident. The ability to freeze cells without killing them seemed to promise relief from the incessant labor required to maintain cell lines that were continuous but not stable enough to guarantee experimental consistency over extended periods of time.

In 1959, at the behest of the Viruses and Cancer panel of the National Advisory Cancer Council of the National Cancer Institute, a Cell Culture Collection Committee headed by Jerome Syverton was formed to "initiate and coordinate a national program for characterizing and preserving animal cell strains and to establish a repository and distribution center for reference cultures."[22] The committee announced the availability of twenty-three strains of animal cells in *Science* magazine in 1964. They wrote that this project was undertaken because of the growing use of cultured cells in research, the instability of cell lines over time, and the need to certify certain cell lines as standard references, "possessing constant and dependable characteristics." Such a program of characterization, storage, and redistribution depended on the "accumulated evidence (i) that cells could be frozen and preserved at extremely low temperatures for long periods of time and (ii) that cells could then be reactivated without significant loss of viability or apparent change in properties."[23] Frozen reference "seed stocks" of cells were placed in the cell repository of the American Type Culture Collection, a government-funded institution in Rockville, Maryland, already set up for the storage and distribution of bacterial cultures. Cell lines were carefully characterized in a number of ways, from chromosome counts to photomicrographs of morphological characteristics of both living and stained cells, and then given an accession number "in the order in which they have been accepted by the committee."[24] HeLa received accession number 2.

The ability to recover cells essentially unchanged, as long as they were frozen at low enough temperatures, rapidly became part of the infrastructure of cell culture, with the American Type Culture Collection cell repository complemented by a variety of other collections. Once again, the scientific community worked closely with the biological supply industry to address their own needs on a large-scale commercial basis. A cell and tumor bank was set up at Microbial Associates in Maryland, by the Cancer Chemotherapy National Service Center. Their frozen repository of about 560 animal tumors and 100 cell lines served as "a valuable reserve of tissue for the various intramural investigators and contractors of the National Cancer Institute . . . making it possible for all investigators to test for potential chemotherapeutic agents with tissue from a common biological source."[25] Within ten years of the successful freezing and thawing of HeLa cells, the storage and distribution of cell lines had been standardized and to a certain degree centralized. For researchers, "the refrigerator may be considered a cold cornucopia that can provide the cell culturist with a virtually unlimited supply of characterized cells."[26]

Work on the freezing of sperm had direct industrial application in cattle breeding. As one freezing researcher put it, the effect of long-term cell freezing had an immediate and profound effect on the temporality of animal reproduction:

Reproduction in mammals normally involves contemporaneous and contiguous action on the part of the two sexes. The advent of artificial insemination abolished this requirement in principle, but in cattle at any rate the practicability of effectively using semen long-stored in the frozen state has enormously extended the scope of the technique. It may be said, in fact, that we have abolished time and space in cattle breeding.[27]

Similarly, frozen cells revolutionized the practice of cell biology by abolishing space and time, or at least some of their effects, from the reproduction of experimental materials. Cells frozen at any particular point in time could serve as their own "permanent record," always available for comparison to derivatives that had been allowed to move farther through time.

The freezing of experimental materials is still an integral part of tissue culture practice; a contemporary introduction to the theory and technique of cell and tissue culture places emphasis on freezing as an integral part of tying together different points in time in an experiment or a set of experiments. The cells themselves have to act as a permanent record of the past of experimentation, because otherwise there is no going back.

> The ability to freeze and preserve cells in liquid nitrogen for many years with minimal loss of viability is one of the advantages of working in vitro. Having a frozen bank of cells provides a backup in case cells are lost due to contamination, carelessness, equipment failure, or a natural disaster. Some types of primary cultures can be prepared in large batches, a large number of vials frozen, and the cells thawed sequentially and studied as secondary cultures so that a large number of experiments can be performed on early passage cells from the same preparation. Alternatively, as one tries to establish a cell line, a few vials of the cells should be frozen every three to five passages to have a permanent record of any changes that may occur with passage number. For normal human cells, which have a limited life span in vitro, expansion and freezing of an early passage bank is the only method that allows similar cells to be used in many different laboratories.[28]

The authors also point out that rigorous labeling discipline is important to the use of freezing, because the cells in a laboratory often outlast the people. "Frozen vials of cells frequently stay in a laboratory long after the student, fellow, or technician who froze them and placed them in the tank is gone. Therefore, a good record-keeping system is an essential part of cell storage."[29]

Thus the 1950s was the decade in which the cloning and freezing essential to contemporary cell culture practices came to mammalian somatic cells; and with both practices, HeLa was often the first and the exemplary test subject, followed by hundreds of other cell lines from animals and humans. Cloning acted to individualize and identify somatic cells as distinct entities, even when they came from the same original tissue. Vegetative, asexual reproduction outside the body became a possible fate for a somatic cell. If it was characterized and frozen, it could live in many different places and times, outlasting and sometimes outliving the scientists who worked with it. The two activities together changed the way scientists worked with cultured cells.

Immortality, Cycle Two

Before the 1950s, both cloning and freezing had been the subjects of intense theoretical debate. In cloning experiments, the individuality and autonomy of the cell was at stake; with freezing, the limits and thresholds of life itself were considered under investigation. Thus both developments shared the characteristic of being profoundly practical solutions to questions that had previously been regarded as fundamental philosophical issues of the nature of life—biological limits inherent to life. When the thoroughly empirical—and, in the case of glycerol, thoroughly accidental—tinkering with apparatus and chemistry showed certain things to be achievable, the boundaries around what one could do to living cells were shown to be manipulable.

The inaugural issue of the journal *Cryobiology,* published in 1964, divided the history of research into freezing living organisms into two periods: 1776–1949 and 1949 to the present. This imbalanced division, hinging on the discovery of glycerol, contrasted the pragmatic concerns of life science researchers wanting to store living experimental material in a stable state for future experimentation with the concerns of the previous period, characterized as being far more metaphysical: "The concept of the threshold of life, lethal point, death temperature, biological zero, and the like was a salient consideration in the theoretical discussion and provided the chief motivation in the study of effects of low temperatures."[30] A. S. Parkes, a member of the laboratory in which the cryoprotective effects of glycerol were discovered, chided earlier researchers for their (failed) work based "on a number of theoretical considerations; by contrast we have proceeded from one empirical observation to another."[31] Similarly, Robert Pollack has commented, in his compilation of classic tissue culture papers, that Sanford, Earle, and Likely showed that the difficulty of culturing single cells was "not philosophical but alimentary."[32]

Freezing and cloning have not lost their significance on a philosophical register. If anything, this manifest ability of the empirical practices of the life sciences to challenge assumed boundaries and intervene so profoundly in living processes generated philosophical perturbation of a sort that we are still trying to come to terms with. This perturbation is evident in the remarkable public life that HeLa has led since 1951. HeLa continues to be used, explained, represented, and narrated in the scientific and popular press, as well as through film and television, making it one of the most storied biological entities of the twentieth century—along with the immortal chicken heart, of course. Indeed, this is the second cycle of immortality as a scientific object, newly refigured by freezing and cloning, through the public imagination. The phenomenon of the public life of biological celebrities has become more prominent in recent years with the high profile of biotechnology and its various poster animals, including OncoMouse, Dolly, and CopyCat. Analysis of the HeLa stories and their relationship to fundamental shifts in scientific practice in the life sciences may help us understand the role these characters play in accounts we tell one another about changing circumstances for life in contemporary scientific culture.

By juxtaposing practical—some might even say mundane—subjects such as nutrient media and deep freezers, with changing public discourses of life and death, I intend to emphasize that these accounts of HeLa are not just "popular" renditions of science but ways that scientists themselves narrate assumption-altering, philosophically disturbing technical change in their practices and objects. It would be utterly artificial to demarcate "scientific" from "popular" narratives within this literature. HeLa was both the first and the standard example of large-scale *in vitro* reproduction of a human specimen that could be everyone's specimen, shipped and mass reproduced, cloned and frozen. It was, literally, living proof of the unexpected autonomy and plasticity of the human somatic cell under appropriate technical conditions. Although the ways that stories are told about HeLa have changed from 1951 to today, reflecting the shifts from the philanthropically funded public health campaigns of the 1950s to commercial biotechnology today, a number of themes remain constant and can be understood as attempts, on the part of both scientists and their publics, to integrate this new form of human biomedical subject, the cell line, into older forms of life stories.

My retelling of the HeLa story is meant to highlight both what has happened between 1951 and today and the thematic constants that thread through these changing stories. First, and most notably, there is the personification of the cell line in the image of the woman from whose body it was established. This is tightly linked to the realization of the new autonomy and individualized identity possible for the somatic cell lineages through mass tissue culture and cloning. Second, the temporal dislocations in assumed forms of life cycles and lineages created by long-term culture and freezing generate the particular forms of immortality in these narratives. Third, there is the slippage between the literary or citational presence of HeLa and its material ubiquity. Where HeLa cells are, and how they are cited, provides material and literary tangibility for the "body" of knowledge being generated in an increasingly large-scale biomedical research enterprise. A constant preoccupation with mass—what she would weigh now—accompanies the discussion of this scientific corpus.

The HeLa Stories

In 1951, when it became clear that HeLa cells were going to continue growing and dividing unperturbed by their artificial environment, the label of "immortality" was applied to them and their role as a cell line quickly overshadowed their use in the cervical cancer study. George Gey distributed samples of HeLa to his colleagues around the world. Because—as one tissue culturist put it— "HeLa cells can be grown by almost anyone capable of trypsinizing cells and transferring them from one tube to another," their cultivation quickly became a widespread practice.[33]

Gey never attempted to patent or otherwise limit the distribution of HeLa cells. He did not anticipate the chain-letter effect of sending out cultures that were then grown up, split into parts, and sent on to others, particularly after their use in the polio campaign. For anyone who did not receive shipments of cell cultures from the Tuskegee Institute, the biological supply company Microbial Associates began growing HeLa cells for commercial sale. In 1954, Gey expressed dismay over the number of laboratories working on HeLa in a letter to a colleague. Gey's correspondent, Charles Pomerat, reacted to this statement with some amusement: "With

regard to your statement . . . of disapproval for a wide exploration
of the HeLa strain, I don't see how you can hope to inhibit progress in this direction since you released the strain so widely that it
now can be purchased commercially. This is a little bit like request-
ing people not to work on the golden hamster!"[34]

George Gey had as little control over the story that he released
into the public domain as he had over the cells. Because of intense
national interest in the subject of polio, the HeLa cell line came to
the attention of journalists very quickly. The National Foundation
for Infantile Paralysis, which raised much of its funding through
public appeal, was very interested in translating the widespread scientific presence of HeLa into a public one. From their point of
view, the story of the cells and their origin was as valuable to the
ongoing effort as the cells themselves. It was human interest that
fueled science, literally, by attracting more donations—of money
or of blood. When the "Director of Scientific Information," Roland
Berg, contacted George Gey in 1953 regarding a magazine story
about HeLa and Henrietta Lacks, Berg completely dismissed Gey's
suggestion that he use a fictitious name for the story.

> It is axiomatic in presenting this type of material to the public
> that to inform them you must also interest them. As one
> who has been writing for the public for the past fifteen years
> in this field, I have learned that you do not engage the atten-
> tion of the reader unless your story has basic human interest
> elements . . . An intrinsic part of this story would be to
> describe how these cells, originally obtained from Henrietta
> Lakes [sic], are being grown and used for the benefit of man-
> kind. Here is a situation where cancer cells—potential de-
> stroyers of human life—have been channeled by medical sci-
> ence to a new, beneficent course, that of aiding the fight

against another disease . . . Thus, in a story such as this, the name of the individual is intrinsic.

"Incidentally," Berg smugly concluded, "the identity of the patient is already a matter of public record inasmuch as newspaper reports have completely identified the individual."[35]

Thus the figure of Henrietta Lacks was brought into public circulation as part of the same economy of national science and philanthropy that brought HeLa cells into laboratories across America. Another journalist writing for *Collier's* in 1954 was more discreet, referring to "an unsung heroine of medicine named Helen L."[36] Helen L. was characterized in this piece as a young Baltimore housewife whose unfortunate early death turned her into an "unsung heroine" because of the HeLa cells' research role. Her death and her immortality were uttered in the same sentence: "Mrs. L. has attained a degree of immortality she never dreamed of when she was alive, and her living tissue may yet play a role in conquering many diseases in addition to the cancer which killed her."[37] The widespread presence of the cells in laboratories was equated with the ongoing, if distributed, presence of the woman's life essence. In this period, the personification of HeLa cells in the figure of Henrietta Lacks was a beneficent one, a story of unfortunate death turned to the benefit of mankind in conquering polio. The cells were understood to be a piece of Henrietta Lacks that went on growing and living, encased in a test tube instead of a body. The cells were seen as universal human cells, and their concomitant personification was in the form of an angelic figure, an immortalized young Baltimore housewife, thrust into a kind of eternal life of which such a woman would never dream.

Throughout the 1950s, the public life of HeLa was very closely linked to the cell line's use in the polio campaign. The National

Foundation for Infantile Paralysis produced a film in the early 1950s called "A New Approach," showing John Enders's tissue culture laboratories and making a point of visually demonstrating the replacement of monkeys as experimental subjects with test tubes of cells. George Gey went on national television in 1951, grasping a tube of HeLa cells, standing in front of a screen on which a time-lapse film of highly magnified HeLa cells moved and divided. Human cells were described as the saviors of the human publics watching them, whether for defeating polio or cancer. After this initial interest, there is little writing about HeLa between 1954 and 1967 that does not take the form of a scientific paper. There are thousands of these: publications about the biochemistry, morphology, behavior, maintenance, chromosomes, metabolism, responses to radiation, cycles, and so on of the HeLa cell. With such settled ubiquity came relative invisibility as a taken-for-granted point of reference. HeLa had become everyone's specimen, and the line began to stand in for a generalized human or cellular subject; titles of these publications sometimes refer to experiments with HeLa but more often of experiments with "mammalian cells" or "the human cell." A factor in this flourishing of research was the availability of living human materials for experiments that could or should not be performed on living persons. HeLa cells were placed near atomic test sites to see the effect of intense radiation on human cells and they went into space; but these were only the most showy of the thousands of experiments done with HeLa.

HeLa was certainly not the only cell line in use; on the contrary, cell culture flourished in this period, with hundreds of other cell lines established. In addition, there was extensive use of "primary cultures"—cells taken from patients and cultured only as long as they were needed. The escalation of toxicity and carcinogenicity testing on cells instead of on animals was one such use of primary cultures. HeLa, however, remained as the most widely used, widely known, and easily available standard cell. This situation has not changed. The number of citations for HeLa cells on MedLine continues to increase; the use of cell lines in general is still increasing.[38]

Immunologist and poet Miroslav Holub writes that HeLa was a result of the "good manners" of research that included the establishment and exchange of cells:

It's just good manners in oncologic or virologic research to try and establish continuous, permanent lines from [malignant cells]. Some comply, but many seek revenge by giving rise to mysterious phenomena caused by the reciprocal contamination of cultures. The most notorious cells of this kind are those of one Henrietta Laks [*sic*], which have been growing *in vitro* as HeLa cells since 1951 and displaying a downright cosmic eagerness and offensiveness.[39]

The notoriety to which Holub refers came only after fifteen years of widespread use. First, new scientific work that studied aging through cell culture revealed that only cancerous cells could keep dividing indefinitely. This result drew a much sharper line of definition between normal and cancerous cells. Carrel's famous immortal chicken heart cell culture had supposedly been composed of normal cells, in which "permanent life" had been induced by removal from the body and manipulation of their environment. This characterization was shown to have been fraudulent in 1961, when Leonard Hayflick demonstrated that normal somatic cells in culture consistently divide for a set number of generations. Cells reproduce by replicating their DNA and dividing into two daughter cells. When a whole population of cells goes through division, it is said to double. Hayflick showed that cells taken from human fetal tissue undergo about fifty doublings. A frozen culture, when thawed, will pick up where it left off and complete about fifty doublings, no matter how long the culture was frozen. Cells taken from adults consistently go through about thirty doublings.[40] What is more, the finite number of doublings is species-specific: chicken cell cultures, for example, go through thirty-five doublings at the most—far shorter than the time the immortal chicken heart "lived." Thus it seemed impossible that Carrel's culture could have been composed of normal chicken cells.

Hayflick concluded that the chick embryo extract preparation Carrel used as nutrient medium provided new viable embryonic cells at each feeding. Others have hypothesized that Carrel's proximity to Peyton Rous at the Rockefeller Institute led to his somatic chicken cells being infected with Rous sarcoma virus and thus rendered capable of the unlimited division seen in cancerous cells.[41] Given the culture was thrown away in 1946, a definitive answer can never be found. More important was the stark distinction drawn between normal body cells and cancer cells. Intrinsic to this distinction was the finite life span of populations of normal cells. Normal somatic cells were euploid—that is, they contained a normal number of chromosomes. Cancer cells were aneuploid, with abnormal chromosome numbers. Immortality was not available to normal, euploid cells except through freezing. They could be "transformed" with a virus or mutagen, but then they became aneuploid and behaved like cancer cells.[42] Immortality was thus a characteristic only of cancerous, aneuploid cells, and it was one of the characteristics that made cancer a menacing and mortal disease of the body.

Malignancy and cancer were already associated with uncontrolled cell proliferation and metasticization; but it was not until after 1966 that HeLa cells were understood or described in these terms. Certainly it was recognized that HeLa cells came from the cancerous tissue that caused Henrietta Lacks's death, but the emphasis had been on their control by scientists, their harnessing as producers of knowledge in the victorious battles against polio, and the less successful but still hopeful attempt to understand and contain cancer. This sense of control came to an abrupt end with the second and more profound disruption of a benign image of *in vitro* immortality.

The ubiquity of HeLa continued, but its invisibility as "standard reference cell" faltered in 1967 with the announcement that HeLa cells had contaminated and overgrown many of the other immortal human cell cultures established in the 1950s and 1960s. Because one human cell looks very much like another, only cross-species contamination—which could be seen by counting chromosomes—had been identified in cell culture up until that time. New biochemical techniques for identification allowed scientists more insight into the cultures they were using. At the Second Decennial Review Conference on Cell Tissue and Organ Culture, geneticist Stanley Gartler announced that he had profiled eighteen human cell lines and judged them all to have been contaminated and overtaken by HeLa cells.

Gartler had tested the eighteen lines electrophoretically for a set of enzymes known to be genetically polymorphous—that is, to differ slightly between different people.[43] All eighteen cell lines contained exactly the same enzyme profiles, indicating that they were actually all the same, rather than eighteen distinct human cell types. All eighteen had the same profile as the HeLa cell. The key piece of evidence in this study was the profile for a particular enzyme called G6PD (glucose-6-phosphate dehydrogenase), which is a factor in red blood cell metabolism. Gartler stood up in front of an audience of tissue culturists and said:

The G6PD variants that concern us are the A (fast) and B (slow) types. The A type has been found only in Negroes . . . The results of our G6PD analyses of these supposedly 18 independently derived human cell lines are that all have the A band . . . I have not been able to ascertain the supposed racial origin of all 18 lines; it is known, however, that at least some of these are from Caucasians, and that at least one, HeLa, is from a Negro.[44]

The terminology of cell culture was already dense with the connotations of lineage, culture, proliferation, population, contamination, and, most recently, malignancy. With the delivery of this paper, Gartler used these terms in a scientific explanation that marked the contaminating cell line as black and the contaminated lines as white.[45]

At this moment, the narratives surrounding the HeLa cell change dramatically. Prior to Gartler's work in 1966, race had not entered into the discussions of either HeLa cells or their donor, Henrietta Lacks. In fact, Gartler had to write to George Gey early in 1966 to ask about Lacks's race. "I am interested in the racial origin of the person from whom your HeLa cell line was initiated. I have checked a number of the early papers describing the development of the HeLa cell line *but have not been able to find any information pertaining to the race of the donor.*"[46] After 1966, the race of the donor was central to the scientific evidence of cell culture contamination, and metaphors and stereotypes of race framed scientific and journalistic accounts of the cell line.

After Gartler made his argument about HeLa contamination, the description of what happened to cells in culture was structured by these metaphors of miscegenation. Scientists passed on this explanation to journalists, who used this narrative to tell the HeLa story to a larger public. The scientists also read the journalists' accounts, footnoting them in their own scientific papers. The warnings about the danger of HeLa contamination, for example, play up a "one-drop of HeLa" theme: "If a non-HeLa culture is contaminated by even a single HeLa cell, that cell culture is doomed. In no time at all, usually unnoticed, HeLa cells will proliferate and take over the culture."[47] One drop was enough.

The racial metaphors altered but did not completely change the way tissue culture had been understood up to this point. Even with the chicken heart cell culture, there was a consistent obsession with hypothetical calculations of the total volume of cells produced by the immortal culture; with HeLa these were calculations of a swamping of a white population by a black one. The calculation is of a fleshliness that not only outweighs the globe but threatens to take it over: "if allowed to grow under optimal cultural conditions, would have taken over the world by this time."[48] The calculation of the putative volume of the culture when "allowed" to multiply freely was not just of a cell culture but also of how much Henrietta Lacks would have weighed, if all her cells were put back together—an "incredible amount."[49]

Gartler's findings and methodology were taken up by Walter Nelson-Rees, director of a cell culture laboratory at the University of Berkeley charged by the National Cancer Institute with keeping stocks of standard reference cells. Starting in 1974, Nelson-Rees began publishing lists of cell lines he judged to be contaminated by HeLa—an alarmingly high number. Contamination proved to be very widespread. It is impossible to estimate how much research was invalidated by the findings that the researchers were mistakenly working on the wrong type of cell. Contaminated cell lines included a set of six cell cultures given to American scientists by Russian scientists under a biomedical information exchange negotiated by Nixon and Brezhnev in 1972.[50]

High-profile incidents such as these, the emphasis on the provenance of cell lines (one of Nelson-Rees's favorite terms), the consistent use of the G6PD marker system, and Nelson-Rees's penchant for personifying HeLa cells all contributed to a revived interest in the figure of Henrietta Lacks in the 1970s and into the 1980s. The inability of scientists to explain why HeLa contaminated other cultures, but rarely the other way around, fed into a characterization of the cells as voracious, aggressive, and malicious. A large number of articles about HeLa and Henrietta Lacks appeared in magazines and newspapers from *Science* to *Rolling Stone* between 1974 and 1977. Unlike the writers in the 1950s, these authors were not interested in the figure of the self-sacrificing housewife. Although cell cultures were being identified by this time by karyological studies—the appearance of their chromosomes—and a number of other systems of genetic polymorphism not characterized as specific to black or white populations, cell identity was still being explained primarily through the G6PD system. HeLa cells were depicted as having a distinct, identifiable biological race due to their particular genetic structure. Michael Rogers, writing in *Rolling Stone,* explained this to his readers in this way: "In life, the HeLa source had been black and female. Even as a single layer of cells in a tissue culture laboratory, she remains so."[51]

This identity as black and female was combined with a character described as "vigorous," "aggressive," "surreptitious," "a monster among the Pyrex," "indefatigable," "undeflatable," "renegade," "catastrophic," and "luxuriant." The narrative of reproduction out of control was linked with promiscuity through references to the cell's wild proliferative tendencies and its "colorful" laboratory life. Michael Rogers reports that he first heard about Henrietta Lacks through graffiti on the wall of the "men's room of a San Francisco medical school library."[52] Nelson-Rees, the self-appointed watchdog of HeLa contamination for the cell culture community, was fond of talking about the appearances of "our lady friend."[53] When describing the letters Nelson-Rees wrote to his fellow biologists when he suspected they were working with HeLa-contaminated cell lines, another journalist wrote, "It was like a note from the school nurse informing the parents that little Darlene had VD."[54] Problems of contamination of cell lines were described as the scientific community's "dirty little family secret."[55]

This pattern shifted again around 1980, when the perception of an economic value for cell lines refocused attention on the circumstances of the cell line's origin in a biopsy for which informed consent was neither asked nor given. The immortalized HeLa/Henrietta Lacks took on a distinctly economic cast in the 1990s. The cell line was perceived by cancer researchers, in one scientist's words, as "the equivalent of the goose that laid the golden egg—a constant supply of precious and essential resource."[56] In media accounts, Henrietta Lacks became a figure of economic exploitation, with a distinctly contemporary right to sue for compensation, personified as the holder of an investment account, where the original capital was those first biopsy cells.[57] These should have had a dollar value from the beginning, because look what they would be worth today, after all these years in the investment account that is the burgeoning biotechnology industry. Her family is cast into the role of the rightful heirs to the proceeds of this "investment" who cannot collect, because nobody ever patented the cells and thus it is difficult to pin down either past or present profit or any one party who is benefiting from the commercial exchange of HeLa cells and all their products and permutations.

Race reenters the story here as demarcating lines of economic power and privilege. As one of George Gey's colleagues commented to him in 1954, it was "out of the goodness" of Gey's heart that HeLa cells, only three years after their establishment, had become "general scientific property."[58] As a black woman from a black family, Henrietta Lacks walked into a clinic at Johns Hopkins where there was no institutional, ethical, or legal framework to ensure that she or her family was in a position to execute any kind of decision—out of the goodness of their hearts or otherwise—as to the fate of the cells. Lacks's family and friends, long left out of the story, began to be interviewed as important players in a drama in which Henrietta Lacks's cells had become important tools of modern medicine without her or her family's knowledge or permission. With contemporary awareness that important tools of modern medicine are also valuable commodities, endless reproduction and worldwide distribution remain part of the story of Lacks's immortality, but the metaphors have become those of the growth of capital while those of miscegenation and contamination have retreated into the background.

Coupled with this sense of economic injustice has come the last addition to the HeLa story: bioethics. Lacks's family has been invited to receive plaques commemorating her contribution to medicine; a conference on cell lines from diverse ethnic populations and a day have been named after her in Atlanta. The language of restitution and overdue recognition accompany these gestures. The story continues to be told again and again. In "Henrietta's Dance," for example, a piece in the *Johns Hopkins Magazine,* the cell line is once again personified as Henrietta Lacks, both by the journalist and by Walter Nelson-Rees, with the same origin story, the same contamination story, but this time a bioethicist is interviewed in order to relate the details to the broader scene of widespread use of human materials in biomedical research today.[59] Indeed the tie between the cell line and the woman's identity are used to link the more general scenario of use of human materials to the notion that this is a possibility for any individual: A tissue sample taken from your body may go on to have an independent, ongoing, and highly distributed life as a biomedical subject.

What She Would Weigh Now

The tenor and content of the HeLa stories has changed over the decades from the 1950s to today, according to changes in both the scientific uses of the cell line and the social context of this scientific work, including attitudes to modern medicine, ideas of biological race, and the economic circumstances of the life sciences. However, they have also remained remarkably constant in several respects, an apparently endless repetition of the same that echoes the proliferative life of the cells themselves. The elements of personification, immortality, imaginations of volume ("what she would weigh now"), and the way HeLa stands in for and makes visible a body of knowledge have remained in place from 1951 to today. How may these elements be understood in relation to the technical accounts of change in circumstances for human cells outlined at the beginning of this chapter? What do personification and immortality narratives have to do with developments in freezing and cloning of somatic cells?

The ever-present slippage between cell line and person is perhaps the most insistent of these characteristics of the HeLa story. In 1970, George Gey died of cancer at the age of 71. Gey's colleagues at Johns Hopkins published a peculiar memorial tribute to him in the journal *Obstetrics and Gynecology*, entitled "After Office Hours: The HeLa Cell and a Reappraisal of Its Origin." They wrote that the original biopsy "secured for the patient, Henrietta

Lacks (Fig 2) as HeLa, an immortality which has now reached 20 years. Will she live forever if nurtured by the hands of future workers? Even now, Henrietta Lacks, first as Henrietta and then as HeLa, has a combined age of 51 years."[60] Beside this statement is the story's Figure 2—a photograph of a young woman, smiling into the camera, hands on hips. Underneath the photograph, the inscription reads "Henrietta Lacks (HeLa)," as if the photograph of the woman held the image of the incipient cell line, as if the woman *was* the cell line that, according to these gynecologists, "if allowed to grow uninhibited under optimal cultural conditions, would have taken over the world by this time."[61] A medical genetics textbook from 1973 similarly uses the picture of Lacks to underscore the person associated with the scientific entity: "medical geneticists making use of the study of cells in place of the whole patient have 'cashed in' on a reservoir of morphological, biochemical, and other information in cell biology derived in no small part from study of the famous cell line cultured from the patient pictured on this page."[62] Here the verb "cashed in" stands for the riches of scientific information—morphological, biochemical, cytological—to be found in the cell. The importance of identification of cell to person was in other words not merely fanciful. In the structure of reasoning behind the use of the cell line, there is an absolute necessity for a link between *in vitro* and *in vivo* life to be maintained; the information gleaned from cells is useless unless it eventually relates back to the biology and then the pathology of the patient. Through the individual patient, the information then becomes applicable to humans in general. The continuity between person and cell line was the rationale for using "cells in place of the whole patient."

With this kind of personification comes the attribution of intention and autonomy, such as Holub's invocation of cell lines that "do not comply," demonstrating an "almost cosmic eagerness and offensiveness." One author waxes poetic, imagining Henrietta Lacks as "HeLa the patient, who has had immortality thrust upon her," as she continues to "wander through medical history as a modern-day Ozymandius."[63] Such a comparison, although resting on a radical misreading of *Ozymandius,* is nonetheless a typical evocation of the spectral quality of the person lurking around the cells. However, imagining the cell line as carrying "her" genetic and chromosomal signatures is not simply an overfired literary interpretation but the recounting of a reconstruction of Henrietta Lacks's genetic profile twenty years after her death: Twenty-five years after her death, "the genotype of the patient Henrietta Lacks from whose cervical carcinoma the HeLa cell was derived was deduced from the phenotypes of her husband and children, and from studies of the HeLa cell."[64]

An insistent personification is thus generated in both scientific and popular texts, by scientists and journalists. As in the case of the immortal chicken heart, discussed in Chapter 2, the practical culture of the laboratory is also the cultural practice through which the biological figure takes shape outside the laboratory. First through polio research, and then through the use of HeLa to figure out all kinds of tissue culture techniques, autonomously living human matter became widely used biomedical research material. More than that, it became possible for the first time for one specimen, taken from one body, to be present simultaneously in thousands of laboratories and thousands of experiments as well as diachronically and repetitively across the lifetimes of the scientists themselves. The possibility of life being removed from the body and never returning to it was contained in this story, an arrow that begins in the point of an individual person and continues without ever looping back. The human body was no longer the only place for the production of new cells. Cloning, with its focus on sameness and vegetative, nonsexual, noncombinatorial reproduction, increased the individuality of the somatic cell because a cell could then have its own distinct lineage and genetic, chromosomal, and phenotypic peculiarities. As a result, this sense of the cell line as an individual being or entity was enhanced, although of course HeLa was in fact composed of all kinds of heterogeneous descendents of the first biopsy tissue. This was a new mode of existence for human matter, and the repetitive insistence on the image of the contained, singular, embodied person who had been Henrietta Lacks as the narrative vehicle for the HeLa cell line was one way to recount this change. It was perhaps necessary to keep the singularity of the one (person) and the multiplicity of the many (cells) together in the same image, in order to grasp the new technical possibilities for mass reproduction of cells and their distribution in space and time.

Entwined with personification is the immortality narrative, in which HeLa and Henrietta Lacks attain immortality "both literally and figuratively," as one writer recalls being told by his biology professor at university.[65] The temporal and spatial dislocation of the body being buried and disappearing from the world but the cells being alive and present in many places in the world is accentuated by the widespread recognition that these cells are often kept in suspended animation in a nitrogen freezer. Accounts of these suspensions of time and the disruption of its recognizable passage in terms of the human life span emerge in uneasy forms: as a memorial tribute to the scientist whom they have outlived; as a modern ghost story in which Henrietta Lacks, "somewhere, with freshly painted toenails and curlers in her hair," dances on, accompanying her cells on the scientific stage;[66] as a form of contemporary unconscious sainthood, like "one of the saints who multiplied in reliquaries after their death."[67] Some people question whether HeLa cells are still human, but no one asks what kind of immortality this is. It is the new form of immortality built into scientific life—disembodied, distributed continuity.

Immortality in the literal sense is also confused with the more abstract notion of having a legacy that "lives on." This legacy, recognized by awards and plaques and the photograph in textbooks, is not just Henrietta Lacks; the cell line also has outlived the scientist who established it, such that its life story substituted for his own in memorials written at his death. For many scientists, HeLa serves as a vehicle or point of reflection for the structure and workings of whole fields or communities of research. According to John Masters, who has edited many books on cell culture in cancer research, HeLa is "the equivalent of the goose that laid the golden egg," yet it is also a story of the failure of peer review, in that the contamination problems first discovered in the 1960s have never been solved with adequate quality control measures or demands that scientists authenticate the identities of their cell lines as part of publication of their results. He estimates that as many as 20 percent of cell lines in use today are mislabeled as something other than HeLa, noting that some cell banks put in fine print that a cell line supposed to be something else may have "HeLa characteristics."[68] In the memorial tribute to George Gey written by his colleagues, the invocation of possible world domination by HeLa was immediately followed by a link drawn between the physical presence of HeLa cells and their literary presence; their physical mass was also a body of knowledge:

> As it is, the mass of HeLa cells that has been grown must be enormous, as is also the information which has been derived from their study . . . In 1968 the Index Medicus responded by

beginning to list publications under HeLa Cell as a separate heading. A HeLa bibliography reads like a Who's Who of cell biology and indeed of modern biomedical science.

HeLa is here a subject heading; it is the "body" of knowledge, a literalization of the networks of association and exchange between scientists. The scale of this exchange, in terms of both time and space, was facilitated both by the "good manners" that dictated informal exchanges of cell lines and by institutions such as the American Type Culture Collection that were set up to facilitate exchanges on a larger scale. The freezer, of course, is a central feature of this exchange network, ensuring the long-term stability necessary to the distributed presence of HeLa.

A constant preoccupation with mass has always accompanied these personifications, immortality narratives, and invocations of HeLa as an embodiment of the scientific knowledge created with the cell line. "What would she weigh now?" people ask—how much more would all the HeLa cells in the world put together weigh now in comparison to the woman's body when she was alive? Why this question? Why this repeated return to physical weight, when it is exactly the potential of any small sample of cells to endlessly make more of themselves that matters, not how many of them currently exist in the world? This question in itself stands as one form of realization of a transition in possibility for human matter: There can be more life in its technological form than in its original, bounded, mortal container.

HYBRIDITY

The standardization and distribution of HeLa cells and tissue culture techniques in the 1950s placed the means of reproduction of somatic cells in laboratories across the world; the development of cell fusion techniques in the 1960s placed the means of recombination of those cells in the hands of scientists. Somatic cells of complex organisms have for most of the years since the rise of cell theory played second fiddle to germ cells, especially in theories about what is inherited from one generation of organisms to the next. For much of the twentieth century, counternarratives of cytoplasmic inheritance struggled in the face of the dominance of the nucleus and its chromosomes and their central role in transmitting heritable material from generation to generation.[1] The egg and the sperm, and their fusion into the embryo, stood as the eternal beginning point, full of potential, while the soma signaled the end of the road of differentiation, when cells got to their fully specialized final states. In the classic Weissman diagram, germ cells continued in a line across the ages, while the poor soma sloughed off at every generation, coming to an undignified and irrevocable end. Questions of how one generation of differentiated somatic cells passed on their differentiated states to the next through mitotic cell division were difficult to analyze.

The possibilities for somatic cell genetics changed with the individualization of the somatic cell through culture techniques; almost as soon as somatic cells in culture acquired the pure lineages conferred by the cell cloning practices developed in the 1950s, achieving their own names and characterized phenotypes, attempts were made to merge and recombine them. As a result, in the 1960s a practice alternately called "cell fusion," "cell hybridization," and "somatic cell mating" emerged. This was an exercise in somatic heredity that was an unprecedented bypass of the germ cells as the sole means of reproduction and recombination of animal genomes. Thus somatic cells also attained at this time what had until then been the province exclusively of the germ cells, particularly in the mammalian world: They too could be starting points, with the status of "parents" or potential candidates for "mating" in the production of novel progeny.

In a complication of the usual model organism relationship, this episode of novel merging of cells resulted from an attempt to physically reshape the biological matter of higher organisms, particularly that of humans, to be more like their simpler models.[2] The earliest experiments in cell fusion were directed at making complex somatic cells live, behave, and exchange genetic material as bacteria did. Much to the surprise of scientists who thought that somatic cells might occasionally exchange a piece of DNA here and there, just as bacteria had recently been shown to do, mixed populations of cultured cells living together in the same dish produced hybrid progeny when two cells of different types fused into one. Once again, the extreme plasticity of biological matter—its ability to adjust and keep living and producing after profound material rearrangement such as the merging of cytoplasms and nuclei—took scientists very much by surprise and opened up an entire field of experimentation. Following up on these spontaneous fusions, over the decade of the 1960s scientists developed methods for the directed, intentional fusion of different kinds of cells, moving gradually from fusing mouse cells together to inducing fusion across species boundaries and then to using cell fusion as a method to juxtapose different biological states, times, and kinds in the same cellular entity.

These fusions probed and proved the plasticity of cells in an unexpected new dimension. During the course of these experiments, biologists first realized that the boundaries of species integrity signaled by infertility, and the boundaries of organismal individuality signaled by immune reactions and rejections of transplanted organs, did not apply to the deep insides of organisms—the interior of cells. In cell fusion, not only the cytoplasms of two cells fuse but also often the nuclei, leading in many cases to a fully functional hybrid cell that can reproduce in culture, sometimes indefinitely. Hybrid cells, it soon became clear, produced hybrid enzymes that functioned perfectly well in the living cell despite their dual-species genetic origin. At its inception, the making of hybrid cells was simultaneously an end-run around the necessity of sexual breeding to attain the recombination central to genetic analysis and a realization of the profound homology between species to be found within the cell.

Making hybrid cells may sound like transgenesis, which is the use of recombinant DNA techniques to cross species boundaries without sexual reproduction. Transgenic animals are made by inserting a foreign gene into an egg, and thereby into the germline of the resulting organism; the adult animal then carries that gene in all of its cells and passes it on to the next generation. Mice, for example, express human proteins if a human-derived genetic sequence is introduced into their germline. Indeed, the rise of recombinant DNA techniques in the early 1970s quickly overshadowed cell fusion. The splash caused by recombinant DNA, as well as by the insistent focus on genetics and molecular biology as the drivers of change in twentieth-century life sciences, has meant that methods such as cell fusion have been marginalized in histories of biology and biotechnology. At best, these histories note that cell fusion is the genealogical origin and practical basis for the making of monoclonal antibodies.[3]

In this chapter, I have little to say about these events "causing" or "leading to" transgenesis or cloning; instead, my focus is on the characterization of an attitude to living matter that was established in the 1960s, marked by the appearance of a new language of cellular manipulation—one of living fragments juxtaposed in artificially reconstituted wholes. This attitude to the living is one that is very much still with us, in part because it was strengthened and universalized by the development in the 1970s of much more specific modes of nonsexual genetic recombination. The notion of the reversibility of biological states, explored using cell fusion in the 1960s, has become central to the idea of "reprogramming" cells with cloning and stem cells in the first decade of the twenty-first century.[4]

As with other moments in which some new and unexpected biological quality or new variety of being has arisen from prodding the cellular lives maintained *in vitro*, we can learn much from these early years in which biological hybridity was being profoundly rethought. Hybridity has had a long history of importance to our thinking about biological difference and its insuperability; Lewis Thomas commented in the early 1980s that the "laboratory trick" of cell fusion seemed to reconfigure ideas and practices of the individuality of living things to an extraordinary degree. He mused that, "in a way, [cell fusion] is the most unbiologic of all phenomena, violating the most fundamental myths of the last century, for it denies the importance of specificity, integrity, and separateness in living things."[5] As with immortality, this is a point at which a concept of broad scientific and cultural salience was altered through its elaboration as a specific characteristic of the lives of cells in culture, after which it carried new and different meanings and practical possibilities. These possibilities struck contemporary observers as startlingly artificial, unnatural, or, to use Thomas's neologism, unbiologic—not making biological sense.

Giving a clean narrative to many topics in the life sciences after World War II is difficult. In cell fusion, events occurred in hundreds of different laboratories in both North America and Europe and involved large numbers of participants, each contributing to a technique that was used for genetic analysis, cancer research, and studies of development. In addition, the uses of the technique eventually moved in directions as diverse as the production of monoclonal antibodies and the creation of the first transgenic organisms.[6] Thus I have organized this story not by the contributions of particular persons or specific achievements but by the concepts and practices through which hybrid entities were generated and by the effects of their experimental use on ideas of biological hybridity.

Three features of the new hybrids are essential to understanding their significance: first, the mobilization of cell fusion as a way to get around the bottleneck of sexual breeding—it was often described as "parasexuality" or "genetics without sex." Second is the direct consequence of genetics without sex—the realization that there were no intracellular mechanisms for recognizing incompatibility between individuals or species. Third is the implications of cell fusion as a medium of juxtaposition of extreme biological difference. As it developed, cell fusion became a technique for experimenting with the reversibility of developmental processes such as differentiation: These mergers allowed the juxtaposition of biological matter across species lines as well as across biological states or times—differentiated/undifferentiated, old/young, and malignant/nonmalignant. Biological time once again became malleable through the manipulation of cells and their medium.

The Parasexual Approach

From the first borrowings by Ross Harrison of his bacteriologist colleagues' protocols and laboratory equipment to Alexis Carrel's confident pronouncement in 1912 that somatic cells from complex organisms could be cultured as easily as microorganisms, tissue culture has in many ways been the expression of a strong, century-long desire to fit complex beings into the same easily manipulable experimental spaces as their simpler single-celled counterparts, bacteria. Not to use bacteria as model organisms for more complex animals, but the reverse: to literally make complex animals more like their model organisms, by making living animal matter conform to the shape, time, and technical forms of simpler experimental models. As I detailed in the previous chapter, half a century of debate over the biological possibility or impossibility of somatic cells living without one another was finally put to rest by developments in the 1940s and 1950s of techniques for coaxing a single isolated animal or human cell to divide to form colonies or strains of cells, all descended vegetatively from the single cell. As a result, single cultured cells gained a status as individuals, as distinct entities with their own lineages, characteristics, and names. They were spoken of as having genotypes and phenotypes.[7] As soon as it was technically feasible to treat somatic cells as individuals, and the spread of tissue culture techniques put these cloned strains within the reach of many scientists, the cells became candidates for genetic studies. In particular, living human material seemed amenable to genetic experimentation as never before. The success of genetic studies in bacteria over the 1950s meant that, once again, microorganisms were looked to, longingly, as models of what one could find out about basic cellular mechanisms, if only the system were simple, controllable, and manipulable enough.

In addition, according to geneticists who had worked with plants and animals, bacterial systems were to be looked at as models for the appropriate length of experiment—that is, hours, not lifetimes. J. B. S. Haldane commented concerning bacterial genetics: "[T]o an old-style geneticist the most striking feature of this work is that re-combination can be studied in an experiment lasting about three hours, as compared with three weeks with *Drosophila melanogaster*, two years with an annual plant, and fifty years for human beings if such experiments with them were possible."[8] The generation time of a human cell in culture was one day, whereas the generation time of a human organism was years long.

One interesting aspect of somatic cell hybridization, then, is that the experimental problem and, to some extent, the appropriate experimental system were absolutely clear, yet the outcome—hybrid somatic cells—was still profoundly startling. The problem from the perspective of midcentury was this: Genetic analysis of higher mammals, particularly of humans, was made difficult by the fact that sexual reproduction took a very long time and was impossible to direct for experimental purposes. The decade of the 1950s saw the discovery and elaboration of what J. B. S. Haldane called "alternatives to sex"—modes of genetic recombination and segregation that happened outside the "usual" mode of genetic exchange. Classic genetic analysis in plants and animals, as writers in the late

1950s repeatedly pointed out, had depended on the "triad of mutation, of fertilization, and of segregation and recombination at meiosis."[9] That is, analysis had depended on the random reassortment of chromosomes that happens when chromosome pairs are split up into the germ cells formed by meiosis and the ensuing recombination when sperm and egg from different individuals fuse at fertilization. The result is progeny with different identifiable characteristics or traits inherited from the parents that can be used to track genetic factors. Of course, bacteria have no such cycle of splitting and merging, but they were shown at this time to transfer or exchange pieces of DNA in a number of different ways. Guido Pontecorvo, a geneticist working at the University of Glasgow, showed that certain fungi had both sexual reproduction and another parallel mode of genetic recombination. The somatic bodies of plants and animals are mostly made up of cells containing two duplicate sets of chromosomes and are therefore referred to as diploid. Most fungi, by contrast, are composed of cells whose nuclei contain a single set of chromosomes (haploid). Pontecorvo discovered that, in the growing filaments of certain fungi, these haploid nuclei would sometimes fuse together to form unstable diploid nuclei. When the cells containing these fused nuclei divided, he found that the daughter cells were different from one another, either because of chromosome loss during division or mitotic crossing-over, which is the exchange of small segments of DNA when chromosomes in diploid nuclei pair up prior to cell division.

The details of these bacterial and fungal systems matter less to this story than the implications read into these modes of genetic change and exchange. In Pontecorvo's own words, "if only one outstanding contribution of microbial genetics to biological thinking had to be singled out, it would be this: the realization that transfer of genetic information from one individual, or cell, to another is not the monopoly of sexual reproduction."[10] Note the crucial invocation of the genetic protagonist here as interchangeably "one individual, or cell," a substitution not possible before either bacterial genetics or the ability to clone single complex somatic cells. Pontecorvo called the system worked out in fungi the "parasexual cycle," a term he coined in 1954 for processes that "bring together in one cell lineage hereditary determinants from separate cell lineages," which thus ensure genetic recombination without sexual reproduction. The etymology of the term, he explained, was meant to indicate biological cycles that "lead to the same end but in a different way."[11] If fungi could have both sexual and parasexual systems, he reasoned, so could higher organisms. "Similar methods of analysis," he wrote, "which by-pass the stumbling-block of sexual reproduction, can be applied—in principle—to the formal genetic analysis of man."[12]

This kind of reasoning was shared by many biologists, including Joshua Lederberg, a central player in elucidating mechanisms of genetic exchange between bacteria; he called repeatedly in talks and conference commentaries in the late 1950s for analogous work to be done with cultured somatic cells: "I am still rather puzzled why biologists show such a strong antisexual bias in the consideration of somatic cells," he chided an audience in 1958, as "every single one of the unit procedures needed for the technical handling of mating has been documented in somatic cells."[13] J. B. S. Haldane, commenting in *The New Biology* on Lederberg and Pontecorvo's work, also singled out tissue culture as the next logical step in doing experiments that would not take lifetimes and did not pass through the channels of sexual reproduction:

Recombination can occur in the absence of a sexual process. This observation may be the key to human genetics. The genes for the Rh antigens and that for elliptical blood corpuscles are carried on the same chromosome. If we could arrange for a man heterozygous at both loci to have 500 children, we could determine their linkage with considerable accuracy. We cannot do this, but we might be able to study his bone marrow cells in tissue culture, to find that they sometimes gave rise to circular blood corpuscles, and to find what antigens these carried. By such techniques it may be possible to map the human chromosomes.[14]

This sort of commentary was accelerated in large part by Theodore Puck's development of cloning techniques, discussed in the previous chapter; these techniques were modeled on and named after techniques for handling bacteria for genetic analysis. References to "cell plating techniques" were part of a language used to describe measures explicitly directed at understanding and using the "mammalian cell as microorganism"; the nutrition studies directed at developing a standardized medium for cells also uncovered some cell lines that seemed to have developed unique nutritional needs during their life in culture. These, as with bacteria, were referred to as "auxotrophic mutants."[15] Puck and Fisher immediately used the new cloning techniques to show that the HeLa population they carried in their lab looked morphologically uniform but contained a mixture of genetically stable cells with differing growth requirements; prior to cloning, there had been no way to separate out this kind of "nutritional mutant."

These were also the years of "the war on cancer" and the attention and funding that this brought to cancer research.[16] Researchers were trying to understand the mechanisms by which some tumor cells became resistant to antitumor drugs, and they too had identified mutants in tissue culture that differed from their parent strains in resistance or susceptibility to certain drugs. For these workers, "the tissue culture cell would appear to be the organism of choice."[17] Like the call for a study of human genetics, tissue culture cells seemed promising as a way to study the genetic basis of cancer, if there was one. Charles Pomerat, inveterate microcinematographer, even suggested in 1958 that, because people were coming around to the idea that phenotypes equal genotypes for mammalian cells, his collection of 250,000 feet of film recording cellular morphology and behavior would provide a convenient archive of such phenotypically readable signs of genetic difference between types of cells. Suddenly, the cultured cell was itself an "organism" with a phenotype that could be read for its underlying distinct genotype.[18] These modes of talking about and isolating cells as genetic individuals with quantifiable characteristic differences in phenotype meant that certain tissue culture practices were easily fit to the microbial model; what remained was to get these cellular individuals to exchange genetic material. The problem, indicated by Pontecorvo's emphasis that in principle it should be possible to do formal genetic studies with human somatic cells, was that somatic cells were not bacteria or yeast or fungi, and the actual method for carrying out the principle was not at all clear.

This analogous reasoning with bacteria motivated biologists Georges Barski, Serge Sorieul, and Francine Cornefort at the Institut Gustave Roussy in Paris to try co-cultivating two strains of mouse cells in the same dish.[19] Although these two strains had come from the same mouse, they had changed enough in culture that they had different abilities to cause tumors when reinjected into mice. In addition, their chromosomes had diverged enough that some of them could be distinguished visually. What the experimenters were looking for was evidence of any transfer of these distinctive properties from one type of cell to another: If small pieces of DNA could move between bacterial cells, they reasoned, then perhaps there would be similar exchanges between somatic cells in culture. What they found was that some of the cells fused and their nuclei fused, creating cells that carried all the chromosomes of the two kinds of cell. These new cells apparently replicated and divided. Although this result was not what Barski, Sorieul, and Cornefort had anticipated, it was clearly a mode of genetic transfer. Barski commented that he did not expect these phenomena to be common; surely metazoan cells, unlike bacterial cells, would have some mechanism that protected "the integrity of their genetic material against external intrusions," and in the body "secure normal and orthodox cell filiation amid mixed cell populations in body tissues."[20]

This work might have languished in obscurity were it not for the avid interest in it shown by Boris Ephrussi, who promptly set out to cause abnormal and unorthodox cell filiation on purpose. Along with Pontecorvo, Lederberg, and Haldane, Ephrussi was sure that somatic cells of higher organisms could and should be used in genetic studies. He repeatedly expressed his feeling that the discoveries and methods of biochemistry and genetics told biologists very little about the operation and differentiation of more complex living cells. Thus he was very explicitly interested in using complex organisms themselves in experiments rather than simply imposing concepts figured out with bacteria on higher animals, opining that "the direct extrapolation from the regulatory mechanisms in bacteria to those of higher animals is regrettably fashionable."[21] As historian of biology Jan Sapp has written, Ephrussi was frustrated by biochemists who treated cells as "bags of enzymes" and geneticists who focused on genes as discrete units acting autonomously from the rest of the cell, dictating its activities, such that "the integrative character of the cell, which is its fundamental property, is bound to escape our notice most of the time."[22] Much like an earlier generation's rejection of a histological cytology as a science of dead entities, this was a refusal of the adequacy of methods that treated living elements as static entities in order to be able to analyze them.

It is not a coincidence that Ephrussi himself had been involved in this earlier era of tissue culture before turning to the genetics of flies and yeast. As with Renato Dulbecco, who returned to the tissue culture he had learned decades before in Italy after years of bacterial virology, Ephrussi practiced tissue culture in the 1920s, in the embryology laboratory of Emmanuel Fauré-Fremiet in Paris. He published a monograph concerning his doctoral work on growth and regeneration in tissue cultures in 1932 before beginning his decades-long work in genetics.[23] In a time when microbiologists and molecular biologists were chastising older tissue culturists for not using tissue culture to solve biological problems in a rigorous enough manner, there was nonetheless a profound need for people with some sensitivity to handling somatic cells in culture to make it possible to use them. Thus, two key figures in the postwar transformation of tissue culture into a practical quantitative tool of virology and genetics had learned the technique many decades before, when the questions being asked of cells in culture were quite different.

Perhaps due in part to this early familiarity with tissue culture, Ephrussi immediately picked up on the earliest signs of cell fusion as one potential technique for doing genetics in mammalian cells. Indeed, historians have argued that Ephrussi's work with somatic cell hybrids is "best understood as a way of transplanting chromosomes, chromosome arms, or blocks of genes into a genetically and cytoplasmically foreign context"—an approach continuous with other modes of transplantation he had been practicing since his early years in the laboratory when he employed a micromanipulator to extract and inject small quantities of cellular substances.[24] Cell fusion was for Ephrussi a variation on transplantation.

Ephrussi and Sorieul confirmed the initial results, claiming the phenomenon was "easily reproducible" with other mixed cultures. Importantly, the results were published in the journal *Nature*, giving them a much wider international readership than the original publication of Barski et al., published in the proceedings of the Paris Academy of Sciences. Ephrussi gradually widened the distance between types of cells merged, fusing cells that had been in culture for a long time with cells taken very recently from an organism; but in general he stuck to fusing different kinds of mouse cells together. Others began to express interest in cell hybridization. John Littlefield at Johns Hopkins University worked on the problem of having some kind of system to select out the "presumed recombinants" from the nonhybrid cells in the mixed cultures; again microbiology provided a model of selective systems, in which one type of cell missing one enzyme was fused with a second type of cell missing another enzyme.[25] When grown on medium lacking the metabolite produced by these enzymes, only the hybrid cells could survive, each partner supplementing the other's deficiency. Still, the biologists were depending on the spontaneous fusion of the cells, and to some extent on the "hybrid vigor" that some of these crosses seemed to produce, because the hybrid cells would often outgrow the "parent" cells even without a selective system. The very invocation of the term "hybrid vigor" in the context of mammalian cells shows that older models of hybridity produced by grafting plants onto one another or by sexual breeding were still formative for the thinking about what kind of entity these cells were.

Often a development that appears to be a decisive break from previous assumptions turns out to contain its own rigid assumptions and unarticulated boundaries. These assumptions are particularly difficult to discern because they are wrapped in the mantle of a departure from past practice. Cell fusion was a surprise—who would have thought that the distinct mammalian diploid cell—that "elementary organism" with its own autonomous powers and double set of chromosomes—could not only merge with another of its kind but also could multiply as a new entity carrying a monstrously large number of chromosomes? Was this not a fundamental shift in understanding the plasticity of somatic cells and their openness to transformations in their genetic content? Indeed it was, but in 1965 a marked break from this early work in cell fusion showed that even those exploring the new area of cell fusion were laboring under the assumption that only some kinds of fusions were possible: For five years, only fusions within species, usually mice, were used in these experiments. Then Henry Harris and John Watkins announced in *Nature* that they had successfully used an inactivated virus to fuse cells from different species, which they described as imposing "a form of artificial sexuality on mammalian tissue cells."[26]

These experiments represented a break from the explorations of cell fusion of the previous five years in three ways. First, they used an external agent, a UV-inactivated virus, to disrupt the cell membranes and force cell fusion, rather than waiting for it to happen spontaneously in mixed populations. The time required to get large numbers of fused cells was thus dramatically reduced. Second, they purposefully chose crosses that reached across species. These experiments, which attempted to cross cells of mice and humans, showed that in the resulting heterokaryons, which contained nuclei from the two different species, both mouse and human genes were being transcribed. The article in *Nature* was illustrated with several microphotographs of cells that showed a single cellular entity, bounded by a single membrane, with two or more nuclei. Third, the sign of continued and combined life that they looked for was the synthesis of DNA, RNA, and protein in the composite cell, which would indicate gene expression in both nuclei. The demonstration that both genomes were active in the same cell departed from the previous emphasis on cell morphology and chromosome identification (karyotyping) for demonstrating hybridity.

These mouse/human cells lived for only days; and despite their manifest activity in terms of DNA and RNA synthesis (perhaps the most meaningful sign of "life" to the geneticist/molecular biologist), they did not appear to do much except survive in their odd new forms for a short period of time. They could not be used to establish a continuous line of cells because they did not appear to divide, and Harris had no way to separate the hybrid cells in the culture from the various parent cells. Despite apparent initial disbelief that cross-species fusions were possible, Ephrussi too published a paper with Mary Weiss later the same year describing the successful fusion of cells across species with the older method of culturing the cells together in the same dish and using a selective system to pick out the surviving hybrids. In this case it was a rat-mouse combination: "the first interspecific cross giving rise to rapidly multiplying mononucleate somatic hybrids."[27] The short-term interspecific hybrid thus became a lineage. In 1966, George Yergenian and Mary Nell combined the two methods to cross the cells of two species of hamsters using inactivated virus, generating a continuous cell line with genetic material from both species.

The confusion concerning the relation between sexual and cell fusion compatibility between species comes through clearly in Yergenian and Nell's work. They used the whole organism as a kind of control for what they were imposing on the cells in the dish. Yergenian and Nell were working with Armenian and Chinese hamsters—their names alone indicated the kind of traditional species boundary that biologists recognized. Geographically isolated for some period of time, these two kinds of hamsters had diverged sufficiently that, when they mated, no offspring were produced. This characteristic was fundamental to the definition of species in the first place. In addition to placing the two kinds of cells in the same dish, they reared Armenian and Chinese hamsters of opposite sexes together, and they used artificial insemination in an effort to induce pregnancies across the two kinds of hamster. Although "natural matings" were observed between the animals, no pregnancies resulted, and artificial insemination resulted in "the formation of defective 2–32 cell embryos."[28] Thus the inflexible boundary of sexual compatibility was tested explicitly alongside the malleable boundary of the cell wall, and the fused cells were explicitly compared to gametes. "In sharp contrast to the expected uniformity of F1 hybrid offspring following the fusion of normal gametes," they wrote, "somatic hybrid derivatives may vary considerably in cellular and clonal morphology."[29] In other words, it was possible to fuse across species and the "offspring" of such crosses could vary widely. Each genetically unique "offspring" could then be used to found a cell line through cloning.

These findings were construed initially in terms of powerful older ideas of the barrier between species based in sexual reproduction. After all, cellular mergers of another sort between species were well known to fail: "The incompatibility between the sperm of one species and the egg of another is well established; in extreme cases an egg fertilized by a sperm of another species immediately expels the nucleus of the sperm."[30] Unlike bacterial systems of genetic exchange, which quickly became labeled with male and female designations for the bacteria in relation to which bacteria gave and which received genetic material, no such names or symbols were applied to somatic cells.[31] There were no protuberances or directions of exchange on which to hang male and female terms, despite the comparison with gamete fusion: It was always the hybridization of two diploid (duplicate) sets of chromosomes into one. Furthermore, unlike bacteria, which were neuter organisms that were not described in terms of sex before the 1950s, mammals already had sexual reproduction and this new kind of recombination had to be differentiated from the meiosis-fertilization cycle. Thus "genetics without sex." Also unlike bacteria, these were manifestly artificial, experimentally produced unions; and fusogens provided the means to impel fusions rather than the experimenters having to wait for them to happen. Meanwhile, the very meaning of "crossing" one organism with another was undergoing fundamental change through these practices.

The utility of these systems for genetic analysis was firmly sealed by Mary Weiss and Howard Green's discovery that many cross-species hybrids, as they continued to divide in culture, would progressively lose chromosomes, usually from only one of the species used in the cross. In mouse-human crosses, in particular, the human chromosomes were preferentially lost, leaving cells with a full complement of mouse chromosomes and only one or two human chromosomes. Fusion produced the necessary recombination. Continuous cell division in hybrids produced the necessary living beings in which to study variously recombined genomes and their expression as RNA and protein products produced by the cell. Then chromosome loss indicated that the system would also have segregation built into it, producing "cells with many different constellations of 'parental' genes."[32] A mouse-human cell with only one human chromosome could be tested for its production of human RNA and proteins, and those molecules could be related back to genes on that single chromosome. "Panels" of cells, each containing a different single human chromosome, could be constructed to start such mapping of human genes—these are still in use today. The use of artificial fusogens such as inactivated virus or, later, polyethylene glycol meant that cellular genetic individuals could be recombined quickly and in large numbers; the cells themselves divided to produce a new generation of cells in only twelve to twenty-four hours. It was, an anonymous editorialist proclaimed in *Nature* in 1969, a "new gift to biology" because, "short of being able to breed mammals as thick and fast as fruit flies, the hybrid cell offers an unrivalled opportunity for genetic analysis."[33]

In this way, somatic cells in culture, including human cells, were brought into the field of genetics, where only whole organisms had been before, and "genetics without sex" became a productive route to the artificial construction of new combinations of living matter. Richard Davidson, an early participant in the field of cell fusion, commented that "somatic cell geneticists have nothing against sex. However, sexual reproduction is not well suited for the genetic analysis of mammals and it is especially unsuited for the genetic analysis of man." Somatic cells provided more complex subjects than the bacteria and bacterial viruses that had been so productive for genetics in the preceding decades. Nonetheless, with the introduction of the technique of cell fusion, the essentials of genetic analysis of complex cells were put in place: "controlled matings, short generation times, and large numbers of progeny."[34] The basic requirements of a genetic system were met: The conjunction of two different genomes and "the loss of chromosomes by successive generations of somatic hybrid cells takes the place of the segregation and recombination that occur in germ cells."[35] In turn, fertilization of the original sperm-egg variety was reclassified as one form of "cell fusion" among others; in half-joking banter during a cell fusion symposium, one scientist commented that he was impressed "by the fact that sperm and their highly differentiated multicellular derivatives, namely experimental cell biologists and biochemists, use similar techniques to get cells together. Both use proteases to clean the cell surfaces, and they both use specific binding components to associate cells."[36]

"Subunits of Both Rat and Mouse"

Perhaps the most surprising thing about hybrids to the scientists working with them was the plain fact of their existence: "the *very fact* that somatic cells of different origins are compatible."[37] Henry Harris, five years after the initial interspecies hybrids were constructed, commented that the most fundamental outcome of the creation of hybrid cells was the general realization of the internal homology of cells:

> In the cells of vertebrates there are, in general, no *intracellular* mechanisms for the recognition of incompatibility similar to those responsible for the recognition and destruction of tissue or organ grafts exchanged between different organs. Not only do the cytoplasms of these different cells fuse amicably together, but their nuclei also; and after nuclear fusion has taken place, the composite cell carries out its functions in a perfectly integrated way, and may, in some cases undergo vigorous and indefinite multiplication.[38]

The gene-mapping functions of somatic cell hybridization were rapidly overshadowed with the rapidity and resolution provided by later developments in genetic sequencing and recombinant DNA techniques of the 1970s and 1980s. However, this fundamental implication of cell fusion—the realization of the internal homology of organisms, on the inside of their cells, so to speak—is still very much with us. Because we have become accustomed to entities such as mice carrying human-derived oncogenes, we may find it difficult to recapture the amazed responses to these first interspecies hybrids. This is only because we live within the state of affairs molded by these very events.

That it took five years from the first cell fusion experiments for anyone to try fusing cells of different species speaks to the strength of the assumption that such a radical cross would not work.[39] In retrospect, however, Harris points out that at the time biology was dominated by:

> increasing knowledge of the extreme specificity of cellular interactions; by the findings of transplantation immunology which showed that grafts exchanged between two individuals, even of the same species, were rejected unless these individual had closely similar genetic constitutions; by the set of ideas encapsulated in Peter Medawar's phrase, "the uniqueness of the individual."[40]

Thus the very practices that made cell fusion possible in the first place—the isolation of the somatic cell as a unique genetic individual—may have strengthened the assumption of its integrity and incompatibility with other individuals/cells. Of course, as I noted earlier, more than a century's worth of theorization on the integrity of animal species had also created a formidable set of both articulated and unarticulated assumptions about the inviolability of certain biological limits.

This episode is one that bears the weight of two questions. First, why didn't anyone try it before? Second, why on earth would it occur to anyone to conduct such an experiment at all? It is entirely possible to have both the forces of possibility and impossibility operating at once. To see how both questions make sense simultaneously, we need to look again at the material conditions that made such an experiment both doable and unthinkable until its manifest success made it obvious to others. It might have been the era of the "uniqueness of the individual," but it was also a time of an intense focus on cells in culture as a means of getting around various biological, temporal, and ethical limitations on experimentation with humans. Tissue culture had reenergized virus research in the 1940s and 1950s; a living human research subject had emerged for both virology and cancer research, and various geneticists and molecular biologists were sure this subject could somehow be of use for them, too. It didn't cost much to try. A telling detail comes out of Harris and Watkins's 1965 paper, when they explain why HeLa cells were selected to be the "human" party in the "mouse-man" merger: "These cells were chosen because they could be obtained in quantity as suspensions of single cells" and because the "nuclei were of two easily distinguishable morphological types."[41] In other words, they were easy to handle as disaggregated single cells in large quantities, and they could be distinguished easily. You could add one part human cells, one part mouse cells, and one part UV-killed virus and come back the next morning to see what had happened. Here we see one of the practical outcomes of the kind of work detailed in the previous two chapters: Cells had become flexible tools, easily accessible, available, and manipulable.

The result—the fused cells continued to live and function—was both the surprising part and the beginning of the longer process of recognizing the fundamental compatibility between cells once you got past the membrane. Boris Ephrussi and Mary Weiss, trying to demonstrate that both the rat and mouse genes in their hybrid cells were actively expressed, looked at the proteins being produced by the hybrids. They found rat and mouse versions of a certain enzyme, which were distinguishable by electrophoresis because of their different sizes, along with three intermediate versions. This particular enzyme, lactate dehydrogenase (LDH), is a tetramer, composed of four subunits. These cells were generating all-mouse enzymes, all-rat enzymes, and interspecific hybrid enzymes generated by the random association of "subunits of both rat and mouse."[42]

This result was a very tangible demonstration of compatibility. Different subunits of an enzyme fit together and worked to perform the regular function of that enzyme in the cell, despite the origin of their parts lying in different species. The example of the rat-mouse enzyme gave much greater specificity to understanding and envisioning the material basis of hybridity by suggesting exactly what was being joined in the merger of two cells and by what mechanism "compatibility" might be working. It was one thing to observe the manifest ability of the cells to continue working—metabolizing, going through the internal rearrangements of cell division. It was another to have some sense of why the genomes of different species could be co-expressed and yet not generate biological chaos, which usually equaled rapid death. These findings were quickly followed by a rush of similar examples of both co-expression and a change in expression of a protein; these results indicated that something present in one genome was able to regulate the expression of something in the other genome once they were thrust together into the same cellular space. Ephrussi and his colleagues, for example, showed that cells that produced pigment in culture would cease producing pigment on fusion, indicating the suppression of one genome by another.[43] Obviously, for this kind of regulation across genomes to happen, the molecular elements of one organism's life processes had to interact intimately with those of the other contained within the same cell membrane.

The same editorialist who had called cell fusion a "new gift to biology" noted that "cell fusion is like fertilization, except that it is an unnatural kind of union that has no right to occur." Elsewhere, there was much repetition of the phrase that these entities were "not mere biological curiosities." This venture to convince others of the seriousness of these entities was somewhat undermined by the publishing of lurid pictures of centaur-like rat-man creatures on the cover of such sober publications as Heterospecific Genome Interaction.[44] These cartoon images show a being with the torso of a man growing out of the body of a rat, then shrinking through two steps so that only the human face, adorned with rat ears, shows out of the rat body. Presumably these pictures were meant to evoke mythological hybrids—centaurs and chimaeras—and the fact that human chromosomes were preferentially lost in crosses with rat cells (see Figure 7).

Reactions to the new hybrids bore the outlines of a surprised realization that things previously thought to be separate and inviolable could be artificially shaped into a new merged form of life that was from the point of intervention ongoing and self-reproducing. Cell fusion was not a normally or naturally occurring process of genetic exchange. Even those processes discovered in bacteria and fungi that had been labeled "parasexual" were ones that scientists had observed in these organisms before they intervened to make them happen. These processes were rare and not the main route of genetic exchange between individual organisms, but they did take place on their own. Although some spontaneous cell fusion had been observed in tissue culture before this period, these observations were sporadic and unsystematic. Still, these observations

HETEROSPECIFIC GENOME INTERACTION

Edited by
VITTORIO DEFENDI

THE
WISTAR INSTITUTE
SYMPOSIUM MONOGRAPH
No. 9

Published by
THE WISTAR INSTITUTE PRESS

were resurrected to serve as a genealogy for the new practice. There was nothing natural about cross-species fusions made by applying UV-killed virus or polyethylene glycol. They were manifestly artificial, they represented a catastrophic intervention, and their existence depended on the surprising ability of the cell to go on living and dividing despite the intensity of this manhandling. In the process, biological hybridity was transformed from a rare phenomenon that sometimes arose from sexual reproduction between species to a fundamental indicator of the compatibility between all organisms, once one got past the cell surface.

The Tissue Culture Cell as Test Tube

In a series of lectures given in 1972 for a general audience, Boris Ephrussi sought to articulate the outcomes of more than a decade of research using cell fusion. At a certain point, he made a fairly bald statement of change in experimental conditions for cell biologists: "The combined use of these facts and techniques [of somatic cell genetics] permits today the production of practically any hybrid one wishes to have for any purpose."[45] In other words, it had become possible for a scientist to make almost any kind of hybrid from across the whole range of biological possibility. He went on to say, "And the upshot of it is that, like real molecular biolo-

Figure 7   The cover of an edited volume on somatic cell genetics shows a rat with a man's head and torso, which gradually becomes more and more ratlike, echoing the finding that rat-human hybrid cells preferentially lose the chromosomes of one species until they are left with a full complement of rodent chromosomes and only one or two human chromosomes. Vittorio Defendi, *Heterospecific Genome Interaction,* 1969. Illustration by James Wilner, used courtesy of The Wistar Institute, Philadelphia, PA.

gists, we now receive with fear each new issue of *PNAS*."[46] Experimenters could hybridize anything with anything else, at the same experimental time scale that "real" molecular biologists worked with—in other words, very quickly. Anyone could make these hybrids, in whatever combination, to whatever end.

The assumptions of incompatibility between species and between individual animals or cells were absolutely confounded by the life of these new hybrid entities. What, therefore, had hybridity become, after cell fusion? To address this question, we must look at how drastic species juxtapositions within the same cell membrane opened the way for other kinds of mergers across other reaches of biological difference. These could be different developmental stages, different levels of differentiation, different pathological states, or different points in the cell cycle. Such fusions constituted an extension of the plasticity of cells in two ways—the compatibility of living substance with itself even across difference, and the ability of the living cell to recover from drastic physical disruption as long as some basic parts were in place. Different juxtapositions of substance made by merging two whole cells together showed the ability of the various substances made by different cells to work together in the same cell to continue life. That the physical act of cell fusion did not destroy or kill the components also inspired a plethora of other experiments concerned with parts of cells rather than whole cells. It was a time of microcells and cybrids, the fusion of various cell fragments containing one or a few chromosomes with intact cells, or intact cells with enucleated cells. The language of the "reconstituted cell" emerged to refer to a viable cell that could function and reproduce even though it was literally rebuilt from living but "nonviable" cell fragments (see Box).

This final section of the chapter explores how both forms of plasticity—compatibility across difference and viability after cata-

Hybrid terminology: A whole series of new words were coined to describe the cell fragments and recombined cells produced in cell fusion.

cybrid: cytoplasmic hybrid obtained by fusing an intact cell with an anucleate cell

heterokaryon: multinucleate cell containing at least two different types of nuclei

homokaryon: multinucleate cell in which all nuclei come from cells of the same genotype and phenotype

minicell/karyoplast: nucleus surrounded by narrow rim of cytoplasm and a plasma membrane

reconstituted cell: cell constructed by the fusion of a nonviable nuclear cell fragment (minicell) with a nonviable cytoplasmic fragment (anucleate cell or cytoplast)

segregation: in hybrid cells, the appearance of new genotypes (or karyotypes) as a result of chromosome losses

synkaryon: mononucleate hybrid cell, derived from somatic cell fusion

Adapted from N. Ringertz and R. Savage, *Cell Hybrids* (New York: Academic Press, 1976).

strophic intervention—became part of the concept and practice of hybridity after cell fusion. It is not possible to detail the outcomes, implications, or further genealogies of each experiment, as they branched off in different directions and faded or spread in use in different proportions. Rather, my intent is to create a picture of the field of experimentation at a moment when all kinds of fusion appeared to practitioners to represent the same approach to the living cell—"the parasexual approach." All these experiments were different means of getting one kind of biological matter from one place or state into intimate contact with another kind, by bypassing the cell membrane without killing the cell, and without passing through the restrictive channels of sexual reproduction.

In 1965, Henry Harris did not rest content with demonstrating the feasibility of cross-species hybrids. He went on to immediately pose the question of what kinds of biological problems can be addressed once you get two nuclei (and therefore two genetic entities) operating inside the same cell membrane, sharing the same cytoplasmic space. From species, he quickly moved to other kinds of biological difference, choosing to cross highly differentiated animal cells such as rabbit macrophages, rat lymphocytes, and hen erythrocytes. All of these cells, particularly the hen red blood cells, were referred to as typical "end cells"—so differentiated in their own specialized functions that they did not synthesize DNA or undergo mitosis. The HeLa cell was crossed with each of these; and in the resulting heterokaryon, the rabbit, rat, or hen nuclei showed DNA synthesis activity. The apparent reactivation of these dormant nuclei by their merger into the same space as a HeLa cell nucleus and cytoplasm demonstrated that the differentiated state characterized by the inability to synthesize DNA was reversible.[47] Again, it did not matter so much to Harris's early purposes how long these hybrids lived or whether their nuclei fused and the cells went on to become self-replicating distinct lines. It was this juxtaposition of very different kinds in one space that really mattered.

In the wake of Monod and Jacob's focus in the 1960s on the regulation of gene expression in bacteria, a great deal of interest was generated around the questions of how genes were turned on and off and how those changes in expression in turn related to the "phenotype" or differentiated state of a cell.[48] What, for example, made a liver cell into what it was, versus a kidney cell? What was turned on, and what was turned off, to produce the differentiated state? Were these states reversible? Obviously, in Harris's experiments something produced by the HeLa cell was triggering DNA synthesis in the previously dormant nuclei of the specialized hen cells. Access to the regulation of genes in somatic cells in any systematic way was not possible before cell fusion, but the level of resolution provided by these experiments was not very high. It was impossible at this stage to say what was regulating what even as the outcome of regulation could be observed. "Genetics," then, encompassed more than heredity in the classic sense of characteristics being passed from one organism to another and genetic analysis done with cell fusion was more than mapping genes onto chromosomes; this new field of study was concerned with the transmission of states of expression from one cell to another. Cells from the same organism but from different tissues could be fused, as could cells of the same species and same tissue that were at different developmental stages (embryonic and adult, for example).

Juxtaposition of these different phenotypic states artificially in the same cytoplasmic space was presumed to equal the juxtaposition of different genetic states. Cell fusion thus seemed to be a tool that scientists could use to dissect the genomes of organisms other than bacteria, particularly the human genome. The word "dissect" was explicitly used in these early writings on the promise of such systems for doing genetic studies, particularly in human cells. Similarly, the study of the cancerous and noncancerous state of cells and the genetic differences between them was, from the very beginning of cell fusion, seen as a key application of the technique. These juxtapositions involved fusion of cancerous and noncancerous cells to create a new hybrid line, followed by reinjection of the hybrids into animals to see whether they caused tumors. The increase or decrease in the ability of the hybrid cells versus the parent cells to cause tumors (their malignancy or oncogenicity) was also seen as a measurable sign of the genetic control of the cancerous state.

These studies in juxtaposing differentiated states or biological kinds eventually led to the production of monoclonal antibodies from "hybridomas," or fusions of B lymphocytes that produced a specific antibody with immortal lymphoma cells in culture. This story has been extensively told elsewhere; and, because of the subsequent utility of an immortal cell line that produces endless amounts of pure antibody, hybridoma technology has been one of the most lasting and prominent results of cell fusion studies.[49] This particular utilitarian genealogy of cell fusion—the directed production of a wanted substance by recombining cells to that end—only became evident as an outcome of cell fusion some years later.[50] At the time, however, these experiments seemed to represent one of the spectrum of uses for cell fusion: an attempt to tease out mechanisms of cellular differentiation within the immune system.[51] How did immune cells specialize to such a degree that they produced a single antibody, and what controlled this production? If two antibody-producing cells were fused, would the hybrid produce one, both, or no antibodies? These kinds of questions, and the resulting juxtapositions of state and kind, led to the specialized character of antibody production being recombined with one of the key characteristics of many cell lines, immortality. In this case immortality was relevant because it meant not endless life of a single cell but endless reproduction of more cells of the same kind. Endless reproduction of a highly productive cell was the outcome.

After the initial flush of success with recombining cells across various types of biological difference, the lack of resolution available through this system began to bother experimenters, particularly those who were trying to shape mammalian tissues to bacterial models. A whole cultured cell had become a much more highly characterized and controlled entity after cloning, but it was still a whole cell, a very complex being containing a great many pieces and substances. After recombination, how could one figure out which thing was responsible for the regulation of one genome by another or the production of a particular protein? Experimenters pursued resolution by merging fragments of living cells. This approach reduced the amount of biological matter, such as the number of chromosomes or the amount of cytoplasm, transferred during a fusion. The ability to create such fragments followed the isolation of a fungal metabolite called *cytochalasin b* in 1964 by S. B. Carter (who created the term by combining the Greek words *cytos*—cell—with *chalasis*—relaxation).[52] In applying cytochalasin b to dishes of cultured mammalian cells, he noted an extraordinary effect: The cells would extrude their nuclei so far that they remained attached to the cell body by a fine strand of membrane. This strand was easily severed with a micropipette or by gentle centrifugation, producing membrane-enclosed nuclei without their cells and membrane-enclosed cells without their nuclei. Carter, commenting that the "potency, reversibility and lack of general toxicity" of cytochalasins made them very good tools for taking apart cells nonsurgically, saw the central significance of the substance he had found: The different cell parts were not destroyed or irrevocably damaged by the process of disaggregation. As long as they were reintegrated into a functioning biological structure within 24 hours, they would continue to live. The maintenance of a membrane meant that the techniques of whole-cell fusion could be applied to these various parts, making them easily incorporable into other cells.

In 1916, Eduard Uhlenhuth had declared that tissue culture had created "a new type of body in which to grow a cell"; in 1974, investigators exploring the uses of cytochalasin b declared that they had generated a "new type of cell" with it.[53] They made what they called "microcells," described as "cell-like" structures containing only a few chromosomes. The enucleated cells left behind—named "cytoplasts" by their creators—could also be used in fusions with whole cells to investigate characteristics transferred by cytoplasmic structures such as mitochondria rather than with nuclear materials. The fusion of cytoplast and cell was referred to as a "cytoplasmic hybrid" or "cybrid," as opposed to the term "hybrid," which was reserved for the fusion of two nucleated cells.[54] With these techniques, scientists tried recombining "young" cytoplasm with "old" nuclei to explore what controlled cellular aging;[55] they determined that resistance to toxic drugs was a phenotypic character that could be transferred in the cytoplasm rather than the nucleus;[56] and they experimented with combining various parts of cells that were in different stages of the cell cycle to find some of the determinants of cell division.

The field of inquiry that emerged around the practices of cell fusion was not bounded by experiments that involved fusing cells with one another, as had been done by Ephrussi and Harris and others in the early and mid-1960s. It was more generally described as "somatic cell genetics"—the new ability to do genetic analysis with entities that had not been amenable previously to the necessary steps of recombination with subsequent segregation. Cell fusion in this context was understood as a central and founding practice that brought the parasexual approach from bacteria and fungi to human and mammalian cell genetics. Pontecorvo explained it as a process that "leads to the same end but in a different way"; and in this case the "end" was moving biological substance from one individual being and recombining it with another one without killing either, making a recombinant entity with qualities of both that could be observed or utilized.[57] Cell fusion showed that things were surprisingly plastic in their ability to recover from their forced interaction and their ability to live together. This end—juxtaposition within the cell body—was the same end for a number of different means that emerged together with cell fusion.

"Cell fusion" initially signaled a range of practices that involved bypassing the cellular membrane without killing the cell, and this term should therefore be read broadly as denoting much more than a single technique. Cell fusion was the coming into being of the somatic cell as an experimental setting for genetics and molecular biology. Already in the 1960s, but increasingly so in the 1970s and 1980s, the intense investigations of DNA, RNA, and proteins produced a huge volume of data about molecular interactions investigated *in vitro*. These *in vitro* results raised the question of how to reconcile findings from disembodied biochemical preparations with what happened inside living cells. Consider this explanation of the growing interest in getting things into cells, which comes from an article concerning early experiments with microinjection. This article appeared in a symposium volume entitled "Cell Fusion," but it describes what would seem to be an entirely different technique—using a micropipette to inject things into a living cell.

The demand for techniques to investigate biologically important macromolecules (DNA, RNA, proteins) within the living cell has increased continually with our growing competence to purify, analyze, and modify them in vitro. Since these molecules are not taken up readily by a cell unless a specific recognition-internalization mechanism exists, ways to bypass cellular membranes efficiently without affecting the viability of a target cell have to be found.[58]

According to these authors, microinjection "turns a tissue culture cell into a test tube." Earlier in the century, the cell in culture had been the *in vitro* to the *in vivo* of the whole organism; now the cell became the *in vivo* to the *in vitro* of biochemistry's investigations of biological molecules extracted entirely from their living settings. Cell fusion was one way to bypass the cellular membrane in the absence of a "specific recognition-internalization mechanism"; microinjection was another means of doing so. The end was the same: a recombined living being in which the effects and mechanisms of particular biological components could be investigated. The cell membrane, rather than the test tube, determined the enclosed space of experiment.

Hans-Jörg Rheinberger has described this development as a turn to the organism as *locus technicus,* where the experiment happens in the wet environment of the cell.[59] Biochemistry had traditionally extracted molecules from organisms and made external representations of them in the experimental setting of the test tube. By contrast, with the advent of recombinant DNA techniques, molecules that had been externally manipulated could be used in "internal representations" in the organism, with the living cell as experimental setting. Rheinberger ascribes this epistemic shift to the ability to edit, copy, and reattach DNA sequences to one another and reinsert them into the cell; these practices grew out of the recombinant DNA work of Cohen and Boyer in the early 1970s. Although the level of resolution afforded by recombinant DNA quickly outstripped what would come to seem the crude fusions of earlier methods, what the story of cell fusion shows is that the stage itself—the inside of the cell as a plastic space of recombination—was emerging simultaneously from other distinct areas of biological experimentation. Another way to put this is that both cell fusion and recombinant DNA were part of a larger shift in ideas and practices of hybridity happening during this period.[60]

Cell fusion was regarded in these early years as one of several routes for doing "gene transfer" outside the channel of sexual reproduction. Many practices described at that time as "gene transfer" were actually whole-scale movements of biological matter in which it was assumed that one of the things being moved along with a nucleus or a cell fragment containing chromosomes or mitochondria was a gene or genes. These various forms of transfer, moreover, were not limited to cells in culture; fused cells could act as the transfer vehicles for genes into whole organisms. One author described this latter approach as extending "the parasexual approach to mammalian organisms."[61] Some of the earliest experiments with transgenic mice, for example, involved cell fusion. In 1978, mouse teratocarcinoma cells were fused with human cells to produce a hybrid line carrying a single human chromosome. Single hybrid cells were then injected into a mouse blastocyst, which was implanted in a pseudopregnant mouse. Some of the resulting offspring, the investigators showed, were chimeras, their body tissues derived both from the blastocyst cells and from the injected hybrid cell. They looked for, and found, signs of gene expression from the single human chromosome carried by the original hybrid cell; then they declared their organism to be an "*in vivo* system for cycling human genetic material through mice" as a way to study human gene expression during mammalian development.[62] Soon after, they published results showing "xenogenic gene expression" in chimeric mice made using rat/mouse hybrid cells, taking the "rat x mouse" multiplication from the level of the cell to the whole organism.[63] Obviously, the making and study of transgenic animals soon departed in other directions, as development of techniques for the direct injection of DNA constructs into the pronuclei of fertilized eggs superseded cell fusion as a mode of transferring biological material from one entity to the other. My point is not that the early use of hybrid cells in making chimeric mice is some kind of single originating point for later transgenesis; rather, these experiments show how the reformulation of hybridity included the emergence of the organism as a biological stage through which very different materials could be "cycled."

Studies of cancer, aging, and differentiation, as well as the methods and consequences of microinjection, transgenesis, nuclear transplantation, and monoclonal antibodies, very quickly differentiated into their own more and less successful or widespread applications. The field of somatic cell genetics was short-lived as a recognizable field with a coherent identity. The journal *Somatic Cell Genetics,* launched in 1976, was renamed *Somatic Cell and Molecular Genetics* less than a decade later, in 1984. By that time, the following critique was already possible, during a discussion of a paper reporting the results of whole- cell fusions:

> Your approach to the project reminds one of the golden age at the beginning of cell fusion, when the field of somatic cell genetics was first initiated. But nowadays one can isolate and modify genes and study their expression after transfer into all kind of cells. Given this continuing progress in molecular biology, what additional insights do you expect to come from your system, where you fuse whole cells together? . . . I am not convinced that you will obtain new insights into the molecular biology of gene expression from such systems.[64]

By 1984, the early days of cell fusion already seemed to be a "golden age," long surpassed by the specificity provided by recombinant DNA.

The New Hybrid

Via the practices of cell fusion, "hybridity" came to mean the conjunction of difference in the same biological space. If fusion or transplantation resulted in ejection or disorganization, it also led to death (and thus no hybrid), whereas interaction and cohabitation resulted in continued life, embodied by a hybrid cell line or organism. In this reconfiguration, hybridity lost the sense of an anomaly or a phenomenon that occurred at the margins of already adjacent categories. Instead, it became embodied at the molecular level—subunits of enzymes, substances expressed by one genome that were "understood" by the components of the other organism thrust into the same cytoplasmic space.

Genetics without sex was the first defining characteristic of the new hybridity. It occurs outside the channels of sexual reproduction, does not concern specialized structures of the reproductive system, and is not limited or defined by the interaction of germ cells. Rather, it is an artificially induced recombination followed by some kind of segregation. At the same time as being artificial, however, it was not designed or planned ahead of time; it happened as one of the accidents of putting living things in a dish together. However intent scientists were on finding some way to do genetics with cultured cells, they were nonetheless surprised by the manifest plasticity shown by cells—their ability to fuse in the dish in a way that they did not in the body. Genetics without sex was more than a way of doing mammalian or human genetics without bothering with the time and impossibility of sexual reproduction in experimental settings. It was an *approach*. The parasexual approach, getting to the same end by different means, encompassed the various ways developed to juxtapose biological difference by combining substance across the cell membrane without death—getting to a hybrid entity that could be used for analysis, whether that was of gene location or gene interaction or the relations of nucleus and cytoplasm.

The parasexual approach resulted in another surprise: Anything could be crossed with anything else. Barriers of species, immunological incompatibility between individuals—in fact, any form of biological difference that could be thought of—fell away one by one as biological matter was fused into new forms that recovered from the drastic physical intervention involved in their making and went on to live, produce, and reproduce. This was the second defining characteristic of the new hybridity: the realization of internal compatibility. These two features, and the manifest artificiality involved in making the new hybrids, produced a new set of experimental objects, words, and kinds of scientists. The reconstituted cell—a cell fragmented into pieces and put back together in a configuration of desired characteristics—became feasible. At the same time, the scientist who had worked exclusively with extracted molecules in test tubes began to learn to work with living cells, as the boundary of experiments shifted to inside the cell membrane.

EPILOGUE:
CELLS THEN
AND NOW

What does the past elaborated in this book have to do with the cell and its milieu in the present? Anthropologists of science and technology have noted the central role of the living cell in Western biomedical and biotechnical settings. In an ethnographic account of amniocentesis in America, Rayna Rapp observes a laboratory in which extracted samples of amniotic fluid taken from pregnant women are put into Petri dishes and cultured for a week or two until the cells in those samples multiply. Once there are enough cells, they are examined microscopically for alterations in chromosome configuration that signal genetic abnormalities such as Down's syndrome. This process has been made routine to such an extent that it is no longer regarded as research; instead, it is a test conducted for the most part by technicians and not scientists. To the outsider, however, it can look utterly strange again: Living parts of human bodies are grown independently of those bodies for a set amount of time to act as proxies for the body and future of the growing fetus. The laboratory is the place where these extracted cells live, growing in fluids that come out of bottles. The ethnographer pauses: "So many of the advances in cytogenetics described above are less than a quarter-century old," she says, and "to an outsider, they often appear to be magic potions." Rapp asks herself, "Who invented and tested each variety of growth media?"—a question shrugged off by her interlocutors, who use things because they work, and that is all they need to know.[1]

Who tested the growth media—a question I address in this book—is in detailed form the question of how cells came to live separately and how they became alienated from their originating bodies. In recounting the contemporary story of patented cell lines made from bodies of indigenous peoples, anthropologist Margaret Lock notes that only recently have living human body parts been anything other than inalienable possessions, those objects that people never trade. "In order for body parts to be made freely available for exchange," she writes, "they must first be conceptualized as thing-like, as non-self and as detachable from the body without causing irreparable loss or damage to the individual or generations to follow." Cell lines are detachable from the body as body parts, and they continue to grow and multiply without the body. Something has made this detachment possible and to some extent normal, dissipating "the mystical or transcendental essence associated with body fluids, organs and tissues."[2]

Elsewhere, patients are literally replaced as sources of research material by thing-like cell lines made from their tissues. Witness the disease-afflicted families whose members donated tissue samples to French researchers at the Center for the Study of Human Polymorphism. This publicly funded institute subsequently became embroiled in controversy over deal-making with a private American biotechnology company to provide access to that donated biological matter for diabetes research.[3] "French DNA" was seen as being sold; but in its most material sense, it was the store of cell lines made from the donated tissue that provided the engine of continuous production of DNA for study and barter. Paul Rabinow, in a reflection on the case of John Moore, who sued for property rights in a cell line originally made from cells taken from his spleen, argues it has become part of our "characteristically late modern environment" that life forms exist such as the "transformed piece of matter from John Moore [that] now lives forever, reduplicating over and over again in jars slowly rotating on racks in a temperature-controlled room in Maryland." The cell line may be ordered for a nominal fee, and the cells are part of a scientific and technological structure in which "these immortalized bits and pieces can be used to pursue more knowledge, produce more health, to yield more profit."[4]

The history detailed in this book complements such studies of the contemporary moment by showing how the living cell has become an alienated, exchangeable technical object of our late modern environment. The aim of historical investigation here is not to show that the origins of things are further back than people think but rather to reduce the particularity of individual instances of biotechnology by showing how kinds of living entities have emerged and will continue to emerge.

In this epilogue, I suggest that the cut taken through twentieth-century biology and biotechnology in this book can contribute to analysis of the contemporary in two primary ways. Both of these concern the questions that humanities or social science disciplines can bring to the life sciences. First, the emphasis on infrastructure, technique, and approach offers a way to organize historical and anthropological research questions that are not organized by novel technical objects in and of themselves. Interest in biotechnology has been increasing in the humanities and social sciences, causing a proliferation of specific case studies of individual technologies or particular processes in specific places. These approaches often take

CULTURING LIFE

the object as given; "embryonic stem cell" or "breast cancer genetic testing" or "gm corn" arrive already delineated as discrete things to study, organized by the logic of the scientific or medical field that produces or uses them. They seem to have solidity and coherence; they demarcate a topic or controversy or activity for observation and study, and to this topic critical theory is then brought. Although local and specific studies are valuable, the approach taken in this book offers ways to complement these analyses, identifying genres of technique and conceptual approach common to biotechnological objects of apparently disparate kinds, thus opening up avenues for research that are not organized by species (especially the human one) or particular object.

A second related contribution concerns the question of biotechnology and the human, often phrased in the basic form, "How is biotechnology changing what it is to be human?" This is not only the purview of the newly flourishing discipline of bioethics. The same query may be posed to anthropological, sociological, literary, historical, and other critical studies concerning life science. What is biotechnology—not as a scientific and technical field but rather as a field of social scientific or critical cultural inquiry into the conduct and conditions of life?[5]

The problematization of human life by biotechnology is best studied by expanding the range of inquiry past the obviously relevant categories of things that impact human lives—objects that are themselves made of human bodies and things that are used therapeutically on human bodies. To limit analysis to these things is to miss much of the operation of biotechnology on the human. The events detailed in the previous five chapters show the movement of life over the twentieth century, from the inside of the organism to outside, an externalized visible form maintained in the laboratory.

The disembodiment of human matter was part of this movement from *in vivo* to *in vitro,* but it was only realized many years into the making of tissue culture as a concept and a practice. Developments with animal matter provided the material and conceptual infra-structure for later experiments with human matter; to this day, the way human cells are cultured has more to do with how cells in general are cultured than how other parts of human bodies are treated. A human cell is handled more like a mouse cell than like a human tissue or organ; for many hybrid cells, the designation "mouse" or "human" is furthermore rather unclear. Human cells cultured in animal-derived media begin to express animal proteins on their surfaces. Were they to be placed in a human body, a human immune response would reject the cells.[6] In terms of the fluid existence of living things in their technical milieu, the category "cell" determines what is done to and with human cells more than the category "human." A human relationship to living matter is es-tablished and made concrete in these practices of transformation.

Biological matter derived from human bodies is a subset of all the biological matter that is out there in the world—it is, in the logic of the life sciences, not endowed with any particularly special qualities other than the usual species variations. Thus the more we develop ways to use embryonic chicken cells, the more we develop approaches to human materiality that are continuous with the way we use chickens. This logic holds for all kinds of organisms: When we change cells, we change what it is to be biological. We are most likely to develop these approaches prior to, or at least in concert with, using human things. Another way to put this is that the usual formula, "biotechnology changes what it is to be human," should have an interim step inserted in order to understand this process of change in any detail: "biotechnology changes what it is to be bio-

223

logical." This interim step, I would suggest, is key to understanding the specificity of "life" after biotechnology rather than "life" after nineteenth-century physiology.

As observers of life science, we are given unreasonably sturdy, highly visible, ready-made categories of the relevance of biotechnology to the human, in part by the burgeoning science popularization industry and in part by the rhetorical underpinnings of funding structures in contemporary life science: health and homology. That is, new developments that use any kind of biological matter are seen as significant to human life, even revolutionary, because they (1) introduce a new therapeutic product, which affects humans by changing their health possibilities or longevity, or (2) suggest that the same is true of human beings and bodies. The latter homology narratives affect people by changing their understanding of human nature via shared evolutionary history and structural or functional homology with other organisms. This change implies the possibility of new processes or information being the same or doing the same thing in human bodies. However, there are less obvious ways in which manipulation of mouse or chicken matter becomes relevant to, or formative for, humans and human matter, such as genres of experiment and material infrastructures for exchanging and storing living matter. Sometimes the freezer matters more than the species, or the medium more than the type of cell cultured in it, in events that become important to how human life is thought about or acted upon. These are just as important as ready-made relevance in understanding where biotechnological change comes from and how it operates.

The now well-rehearsed set of conversations around the cloning of adult organisms in the late 1990s is a case in point. A new reading of cloning is done here as a demonstration of what this book's approach can offer to studies of the contemporary moment, and how it reorients questions of biotechnology and the human. As Sarah Franklin has noted, the event that was Dolly has ironically enough produced an incredible proliferation of sameness in terms of responses about the so-called ethical or cultural aspects of cloning.[7] Will parents clone their deceased children? Will adult nuclear donors be the twin-parent-ogre of nuclear recipients' lives? Where will all those enucleated eggs come from, anyway? Because it raised these questions, cloning was given as evidence of the thunderbolt of the new power of biotechnology hitting human existence at its core, helped along in no small measure by pronouncements by the scientists themselves of this experiment as "The Second Creation" and so on. But what if we refuse the pressure, and don't make the leap directly from nuclear transfer to human nature? I suggest this event be read not as one that foreshadows the ability to clone humans, human organs, or even transgenic sheep producing human blood clotting factor in their milk. Rather, it is a tale of cell science and its attendant manipulations, which alters what it is to be made of cellular biological matter—a change that is very much still pertinent to the present and the imminent future.

Before There Was Dolly There Was Frostie

Much has been made of the Dolly experiment as the unprecedented creation of a new individual from the nucleus of an adult cell, proving that differentiation and perhaps ageing are not final as well as the scientific and economic possibility of creating more identical organisms from individual, adult organisms. But these are not mandatory as terms of discussion just because they are prevalent. Instead, cloning adult organisms may be retold as a tale of cryobiology and cell synchrony. Cryobiology, as described in Chapter 4, is the science of freezing living things such that they are still alive when thawed. It is standard practice these days to freeze microbes, cells, and embryos for later use. And how does one synchronize cells? Synchronization refers to the practice of forcing each individual cell in a population of cells that are growing in a culture dish to go through the stages of the cell cycle—growth, DNA synthesis, and cell division—at the same time. Cells would not go through stages simultaneously unless the cells were deprived of growth factors or subjected to various other insults. If one deprives a whole population of some nutrient required for a stage of the cell cycle, all the cells in the population will stop at that point. Adding the withheld substance will then cause them to all divide simultaneously, on cue.

As detailed in Chapter 4, in 1949, Chris Polge and his colleagues at the National Institute for Medical Research in England accidentally discovered that glycerol protected sperm cooled slowly to below freezing, with the result that the sperm was still alive after thawing again. This result spurred a flurry of research into adding things to the cellular medium as cells froze; and before the 1950s were out, red blood cells, ovarian tissue, sperm, and cell cultures were being frozen and thawed with good survival rates. Frozen semen was a boon to the agricultural artificial insemination business, as it would be later to reproductive biology more generally.[8] The ability to suspend and transport frozen cells meant much greater spatial and temporal flexibility for disembodied living cells. The wider community of biologists using cell culture also benefited because cell lines could be grown up, frozen, shipped, banked at a central location, referred to later, and preserved unchanged. The central storage of cell lines at the American Type Culture Collection dates from the early 1960s. To keep large numbers of cell lines going by continuous culture, without outside contamination, over decades, was an unsupportable task; the freezer therefore acted as a central mechanism both within individual laboratories or companies and within the biological research community more generally to stabilize and standardize living research objects that were by their nature in constant flux.

In 1983, Ian Wilmut was involved in work that produced the first healthy calf raised from an embryo that had been frozen. In retrospect, Frostie garnered ever so little media attention, but as an experiment may be understood as continuous with the initial explorations of freezing in the 1950s and the more famous cloning experiments of recent years in two ways: at the level of material practice, due to the role that freezing plays, and at the level of something like genre—a genre of experiment directed at the controlled stopping and starting of biological time. In the case of the experiment that led to Dolly, tissue from the udder of a pregnant ewe that had been frozen since 1995 in a separate institute (the Hannah) was brought over to the Roslin Institute, thawed, cultured (which means the cells were plated on a Petri dish and bathed in nutrient medium so they would start growing and dividing again), and used as "nuclear donors."

That the cells had been frozen for a few years wasn't particularly significant to the main point of the experiment: the demonstration that the nucleus of an adult differentiated cell could be used to clone a whole new individual. But how, once Dolly had been born and was conveniently continuing to live despite the many insults visited upon her originating cells, could the scientists tell that she was "genetically identical" to the adult ewe from which the transferred nucleus had come, given that the ewe was long dead? The scientists went back to the freezer and got out another piece of the tissue they had used to start the cultures of nuclear donor cells in order to make the comparison. All the disassembled generations, the novel simultaneities, the gaps of time between death of one generation and birth of another with a suspension of continuity between them—all of these deeply unsettling temporal disruptions depend to some degree on the rather banal presence of a working deep freeze. This piece of equipment is now incredibly commonplace, though it was not fifty years before. Thus the story of making a cloned sheep not only suggests the possibility of cloning humans but also underlines the conditions of its own possibility: the ability to freeze, halt, or suspend life and then reanimate, as an infrastructural element of contemporary biotechnology. In short, to be biological, alive, and cellular also means (at present) to be a potential "age chimaera," to be suspendable, interruptible, storable, and freezable in parts.[9]

Experiments in cell synchrony also disrupted any sense of biological cycles being inevitable, fixed in duration, or imperturbable. Cell synchrony was noted in dividing marine eggs for more than a century. The fertilized egg divides into two, and then the two cells both divide simultaneously to make four cells, and then the four cells divide simultaneously to make eight, until a certain level of multicellularity is reached and the cells start dividing at different rates. In the 1950s, scientists working with simple single-cell organisms such as microbes and amoebae, which could be kept in populations in the laboratory, realized that a similar synchronization of division in all the cells in a population could be artificially induced by exposing the cultures to cycles of light and dark or raising or lowering the temperature sharply.

In the very early 1950s, periodic DNA synthesis was of unclear significance; with the movement of DNA to the center of biologists' attention as the hereditary material, it became much more clearly important to understand the process by which one cell became two with two sets of chromosomes. The investigation of the cell cycle, as it came to be called, was the dawning realization that cells were not just "resting" between divisions, as had been previously assumed, but instead the DNA in the cell was undergoing various changes as the cell synthesized another copy of its chromosomes in preparation for division. These various steps are marked by biochemical changes—one could not investigate these changes in levels of enzymes or the increasing volume of DNA synthesis if all the cells in a test population were at different stages in the cell cycle. Furthermore, these things could not be measured in single cells; they had to be "amplified" by making the same thing happen in lots of cells at once, so the molecules involved could be measured. Thus, "as an experimental technique, cell synchrony was developed primarily for the amplification of time-limited events within the cell cycle."[10]

The other major figure in the cloning story, Keith Campbell, had for a good part of his career worked on cell cycle research in yeast and frogs, meaning that he was very good at the "amplification of time-limited events within the cell cycle." He transferred this expertise to working with mammalian cells. Once the mammary gland cells had been thawed, they were cultured, and synchronized by withholding growth factors from the culture. Thus the nuclei used as "donors" to put in the enucleated eggs were taken from the cells when they were all at a particular point in the cell cycle. Campbell and Wilmut claim that the age of the organism the cells came from—or the number of times the cells divided in culture—or the degree of differentiation of the cell—does not matter as much as catching the nucleus of the cell at this particular point in the cell cycle.

It may not matter whether a donor cell comes from a young embryo, a fetus, or an adult animal, or whether it is cultured before transfer or not, or, if it is cultured, whether it goes through one passage or a dozen or more—that is, it may not matter as much as everyone anticipated. If you adjust the cell cycles of donor karyoplast and recipient cytoplast, you can produce viable reconstructed embryos from differentiated cells and perhaps, with better understanding and technique, from any kind of cell.[11]

They also manipulated the enucleated egg. Before injecting the cycle-adjusted nucleus the egg was put in calcium-free medium. An egg penetrated by a sperm is "activated,"—in other words, it is induced to enter the series of steps in which the fertilized egg divides to become an embryo by the accompanying inrush of calcium. Normally, therefore, puncturing the egg to remove its chromosomes and inject a nucleus will cause activation. Withholding calcium acts to delay activation—allowing for the adjustment of the cell cycle of the "recipient cytoplast." The scientist may thus manipulate the egg by adding or withholding calcium in the medium, and both the giving and the receiving cell are kept poised in certain temporal states amenable to the wanted outcome. In other words, the matter of the cells is manipulated, but in addition great effort is put into controlling how they live in time.

Importantly, the scientists put as much emphasis on the technique as on its refutation of what everyone thought would matter. Wilmut et al. indicate that a choice has been made, between an idea of the immovable intrinsic age of living matter (according to whether it comes from an adult or has been living in the laboratory for a long time), and an idea of biological time not as a boundary but a moveable—plastic—quality. This is an attitude to living matter more likely to be found in individuals who have spent decades freezing and thawing, stopping and starting cell cycles than in other kinds of life scientists.

By the time it was put to use in this particular experiment, cell synchrony had become a relatively familiar technique of cell biology, as had the idea that cells have cycles. As had the various technical aspects of getting cells to live outside of the body where they can be experimented on and observed, and the media in which they live can be fully controlled. As had the practices of extracting cells from complex bodies and those of manipulating reproductive cells outside of the body before reimplantation in the body. The authors themselves admit that, like so many other events in science, the novel combination of existing techniques resulted in something startling to all. Pointing out that the experiment was constituted by these techniques of cell culture and cryobiology and cell synchrony is not to say that it was all old hat and that the actual origin of their work lies elsewhere. There is no need to argue the long-term importance of this event as its aftermath is still unfolding. However, in order to pull the terms of analysis away from claims of revolution in this single development, it is necessary to move away from the particularity of this event—to see cloning as an extension of an infrastructure that has been in the making for the better part of a century. This in turn is an effort to find a way to speak to its significance as part of (and not the cause of) the ongoing operationalization of biological time—not just its suggestion of the possibility of doing the same procedure in humans.

These practices, now standard in contemporary biology and biotechnology, are also standard in that they assume and exploit a certain plasticity of organisms. That is, the ability of living things to go on living, synthesizing proteins, moving, reproducing, and so on, despite catastrophic interference in their constitution, environ-

CULTURING LIFE

ment, or form. The very ability to grow cells outside of bodies in artificial environments or on scaffolds, to puncture eggs and inject foreign things into them, to cut and paste genetic material without killing the organism in question altogether, are also good examples of this plasticity. Where would biotechnology be, if after being spliced or frozen or fused or extracted from its original environment, the cell or organism just up and died? In my view, the history of biotechnology from 1900 to now may be described as the increasing realization and exploration of the plasticity of living matter. As with cryobiology and cell synchrony, the manipulation of the plastic matter of the organism is often, if not inevitably, linked to a disruption of temporality, whether that be of lifespan or continuity or smaller scale cycles of growth and metabolism. Whether halting something in a certain state—for example, inducing stem cells to remain in a state of continuous potentiality as if they were blastocysts for eternity, or driving something to completion like a transgenic salmon—material interventions result in things living differently in time.

Biotechnology Changes What It Is to Be Biological

As a subset of this longer twentieth-century course of biotechnology, the cloning story makes it matter differently to be composed of cells and cell cycles. Being a cellular entity after cryobiology and cell synchrony means being freezable and open to artificial synchronization; any living thing made of cells, after these interventions, becomes an object that can be stopped and started, suspended and accelerated. Being cellular after cloning entails a different sense of biology and time: What is lost is the assumption of biological progression being yoked to historical time in any given, predictable way. The operationalization of biological time is

232

a dominant characteristic of the interactions of humans and cells in technical environments over the last fifty years. This formalization of techniques of plasticity and temporality has been enabled by the infrastructural buildup of freezing technologies and cell-cycle interventions and concepts (which have reached the status of unarticulated assumptions). In short, living matter is now assumed to be stuff that can be stopped and started at will. It is these changes that are at work in the production of novel cellular objects today, of which cloned sheep or "reprogrammed cells" are but single examples.

Methodologically, then, the approach via technique reveals different, previously invisible modes of connection between the yeast, chicken cells, and other life forms of twentieth-century biology, and the conduct and questions of human life. It introduces a specificity to understanding how biotechnology, with its characteristic interventions in plasticity and temporality, changes what it is—what it means at any given moment—to be cellular living matter. As an approach to the living, biotechnology changes what it is to be biological, a step that must be analyzed before leaping straight into how biotechnology changes what it is to be human.

Detailing this step provides different avenues of analysis than the accepted links of therapeutic applicability or genetic homology, and gives the observer of the life sciences a way to cut across the structure of arguments and terms of debate already well defined by other agendas and actors in the currently very public life of biotechnologies. This is as true of any specific example one wants to pick as it is of cloning, which is something to think about at a time when it seems like every second news item is about embryonic stem cells. Just as gene therapy and cloning have come and then diminished as high-profile scientific objects or processes, so will embryonic stem cells settle from their current prominence; but the conditions that produced all of these novel forms and objects will still be in operation, busily generating yet more new things and perturbing human biographical narratives.[12]

I should point out that this criticism of an exclusive focus on human things for study is sympathetic with, but is not the same as, the call first put forth by Bruno Latour to include nonhuman actors in stories of the history and sociology of science, as a means of avoiding the assumption that all scientific or technical change springs from human thought. Human genes are also nonhuman actors in his sense, as they are not (usually) endowed with thought. I wish to specify only that human biology and medicine are a subset of how the life sciences function, and are therefore not representative of the vast amount of nonhuman biological work in the past and present that forms the possibilities and concepts of action with human matter today. The exclusive attention to things human is a difficulty that is particularly acute in the study of modern life science because of its scale and diversity, and the aforementioned temporal feature of its reasoning in which the human application comes late in the process.

Keeping an eye on practice, protocols, methods, technique, touch, or infrastructure provides access to the ways in which work on some life (nematodes, insects, yeast) reshapes human life by introducing systematic change into biological existence. Perhaps most importantly, this methodological focus on genres of technique and infrastructures of research allows room for the vast realms of contemporary biological practice and biotechnological intervention that are not based directly on human matter or health or reproduction. We thus gain better access to the question of how, specifically, altering any kind of biology—yeast, fruit fly, nematode, slime mold—is to alter what it is to be biological, without having to assume that their cultural significance derives only from their direct, one-to-one relation to human health and wealth. Gillian Beer writes of Darwin's *Origin of Species by Means of Natural Selection,* as a text that spoke of "survival and descent, extinction and forgetfulness, being briefly alive and struggling to stay so, living in an environment composed of multiple other needs, coupling and continuing, ceasing to be: all these pressures, desires and fears are alerted in this work without any particular attention being granted the human person."[13] Similarly, the contemporary texts of biotechnology, particularly the materials and methods sections of its thousands of constitutive publications, speak of mutation and revision, senescence and parasitism, multiplicity and infinitude, living in an environment composed of multiple other technologies, profit and proliferation, totipotency and replication, also without necessarily granting centrality to the human person. Certainly the focus is often on the implication for human health, and this is accentuated by the funding and investment structures of contemporary research. However, must the observer of the life sciences exclusively focus on how biotechnology is always directly about human biology and human nature? My argument is that to do so is paradoxically to lose sight of much of its power in contemporary culture. Once we have a more specific grasp on how altering biology changes what it is to be biological, we may be more prepared to answer the social questions that biotechnology is raising: What is the social and cultural task of being biological entities—being simultaneously biological things and human persons—when "the biological" is fundamentally plastic?

NOTES

INDEX

NOTES

Epigraphs: H. G. Wells, "The Limits of Individual Plasticity," in R. M. Philmus and D. Y. Hughes, eds., *H. G. Wells: Early Writings in Science and Science Fiction* (Berkeley: University of California Press, 1975), pp. 36–39; Alexis Carrel, "The New Cytology," *Science* 73 (1931): 297–303.

Introduction

1. Quoted in Philip J. Pauly, *Controlling Life: Jacques Loeb and the Engineering Ideal in Biology* (Berkeley: California University Press, 1990), p. 5.

2. Sally Smith Hughes, "Making Dollars out of DNA: The First Major Patent in Biotechnology and the Commercialization of Molecular Biology, 1974–1980," *Isis* 92 (2001): 541–575.

3. Martin Kenney, *Biotechnology: The University-Industry Complex* (New Haven: Yale University Press, 1986).

4. Robert Bud, *The Uses of Life: A History of Biotechnology* (Cambridge: Cambridge University Press, 1993).

5. Bronwyn Parry, *Trading the Genome: Investigating the Commodification of Bio-Information* (New York: Columbia University Press, 2004).

6. Rayna Rapp, *Testing Women, Testing the Fetus: The Social Impact of Amniocentesis in America* (New York: Routledge, 2000). Margaret Lock, "The Alienation of Body Tissue and the Biopolitics of Immortalized Cell Lines," in Nancy Scheper-Hughes and Loic Wacquant, eds., *Commodifying Bodies* (London: Sage Publications, 2003), pp. 63–92. Paul Rabinow, "Severing the Ties: Fragmentation and Dignity in

Late Modernity," *Essays on the Anthropology of Reason* (Princeton: Princeton University Press, 1996), pp. 129–152. Hannah Landecker, "Between Beneficence and Chattel: The Human Biological in Law and Science," *Science in Context* 12 (1999): 203–225.

7. A history of cell biology would be a different thing, as cell biology as a distinct subfield of the life sciences came into being only after World War II, through the direct campaigning of individuals such as Keith Porter to name the field "cell biology" and to develop professional organs such as societies and journals to cater to this particular field. See Carol L. Moberg, "Keith Porter and the Founding of the Tissue Culture Association: A Fiftieth Anniversary Tribute, 1946–1996," *In Vitro Cell and Developmental Biology* 32 (1996): 663–669.

8. Critics of gene-centrism from inside and outside science have of course been vocal for many years. The swing away from the DNA coding sequence as the be-all and end-all of knowing how life works can be seen in many current aspects of biology such as the growth of research on RNA and its unexpectedly complex role in controlling gene expression; the field of evolutionary developmental biology (evo-devo) and its detailing of the complexity of phenotypic variation compared to genetic variation (John Gerhart and Marc Kirschner, *Cells, Embryos and Evolution: Toward a Cellular and Molecular Understanding of Phenotypic Variation and Evolutionary Adaptability* (Malden, MA: Blackwell Publishers 1997)); the growing list of epigenetic, non-DNA traits inherited through cell division (Eva Jablonka and Marion Lamb, *Evolution in Four Dimensions: Genetic, Epigenetic, Behavioral and Symbolic Variation in the History of Life* (Cambridge: MIT Press, 2005)); and the study of the spatial and temporal organization of cellular membrane systems in life processes (Lenny Moss, *What Genes Can't Do* (Cambridge: MIT Press, 2002)). These fields of study have also begun to disturb the popular discourse of genetic determinism; the same science journalist who declared in a 1999 bestseller that "the idea of the genome as a book is not, strictly speaking, even a metaphor. It is literally true" in 2003 was writing in his next bestseller that genes are not "puppet masters or blueprints. They are both cause and consequence of our actions." Matt Ridley, *Genome: The Autobiography of a Species in 23 Chapters* (Harper Perennial, 1999), quotation, p. 6, and *Nature via Nurture: Genes, Experience, and What Makes Us Human* (HarperCollins, 2003), quotation, p. 6. For an anthropological analysis of this shift, see Margaret Lock, "The Eclipse of the Gene and the Return of Divination," *Current Anthropology* 46 (2005): 547–570.

9. Sarah Franklin and Margaret Lock, eds., *Remaking Life and Death: Towards an Anthropology of the Biosciences* (Santa Fe: School of American Research, 2003), p. 13.

10. Paul Nurse, a cell biologist and Nobel laureate who is now president of Rockefeller University, was asked in 2003 what fact he found most impressive about the cell, and he answered: "That many of the interesting properties of life are found in a single cell. The cell is the basic structural and functional unit of life. It is life itself. It is chemistry made into biology. That's what a cell does. The basic processes in a cell are simply chemical reactions, but organized in such a way that we get biological behavior, behavior like life, purposeful behavior—the ability to produce yourself, the ability to organize yourself. I am talking about 'myself' as a cell." "A Conversation with Paul Nurse; New Rockefeller Chief Discovered Lessons of Life in a Cell of Yeast," *New York Times*, May 13, 2003, F-2.

11. Jane Maienschein, *Whose View of Life?: Embryos, Cloning and Stem Cells* (Cambridge: Harvard University Press, 2004); T. H. Morgan, "Regeneration of Tissue Composed of Parts of Two Species," *Biological Bulletin* 1 (1899): 7–14.

12. Sarah Franklin, "Stem Cells R Us: Emergent Life Forms and the Global Biological," in Aihwa Ong and Stephen Collier, eds., *Global Assemblages: Technology, Politics and Ethics as Anthropological Problems* (Malden, MA: Blackwell Publishing, 2005), p. 60.

13. Michel Foucault credits Canguilhem with the recognition of the notion of "recurrent history": there are "several pasts, several forms of connexion, several networks of determination, several teleologies, for one and the same science, as its present undergoes change: thus historical descriptions are necessarily ordered by the present state of knowledge, they increase with every transformation and never cease, in turn, to break with themselves." Michel Foucault, *The Archaeology of Knowledge* (London: Tavistock, 1972), p. 5, quoted in Nikolas Rose, "Life, Reason and History: Reading Georges Canguilhem Today," *Economy and Society* 27 (1998): 154.

14. For literature on the history of cell theory and biology of the cell in the nineteenth century, see Georges Canguilhem, *A Vital Rationalist: Selected Writings from George Canguilhem*, ed. François Delaporte, trans. Arthur Goldhammer (New York: Zone Books, 1994). Marc Klein, *Histoire des origines de la théorie cellulaire* (Paris: Hermann & Cie, 1936), Marc Klein, *À la recherche de l'unite élémentaire des organismes vivants: histoire de la théorie cellulaire* (Paris: Palais de la Decouverte, 1959). Arthur Hughes, *A History of Cytology* (London: Abelard-Schuman, 1959). François Duchesneau, *Genèse de la théorie cellulaire* (Paris: Vrin, 1987). Paul Weindling, "Theories of the Cell State in Imperial Germany," in C. Webster, ed., *Biology, Medicine, and Society 1840–1940* (Cambridge: Cambridge University Press, 1981), pp. 199–155. Paul Weindling, *Darwinism and Social Darwinism in Imperial Germany: The Contribu-*

tion of the Cell Biologist Oscar Hertwig (1849–1922) (Stuttgart: Gustav Fischer Verlag, 1991). Laura Otis, *Membranes: Metaphors of Invasion in Nineteenth-Century Literature, Science, and Politics* (Baltimore: Johns Hopkins University Press, 1999).

15. Examples of surveys that concentrate on the history of genetics in order to tell the history of twentieth-century biology include Peter Bowler and Iwan R. Morus, *Making Modern Science: A Historical Survey* (Chicago: University of Chicago Press, 2005), Garland Allen, *Life Science in the Twentieth Century* (New York: John Wiley & Sons, 1975). There are of course some exceptions. See Jan Sapp, *Genesis: The Evolution of Biology* (Oxford: Oxford University Press, 2003), a survey of the history of biology whose atypical inclusion of the biology of cells in the twentieth century is due no doubt to Sapp's previous scholarship on the history of cells and cytoplasm as an important counternarrative within twentieth-century genetics. Jan Sapp, *Beyond the Gene: Cytoplasmic Inheritance and the Struggle for Authority in Genetics* (Oxford: Oxford University Press, 1987).

16. François Jacob, *The Logic of Life: A History of Heredity*, trans. Betty E. Spillman (Princeton: Princeton University Press, 1973 [1970]).

17. H. G. Wells, *The Island of Dr. Moreau* (New York: Bantam Classics, 1994 [1896]).

18. Pauly, *Controlling Life.*

19. Hans-Jörg Rheinberger has noted that many biological entities balance somewhere between their reality as living things independent of our manipulations and, conversely, their existence as deliberately constructed things. They have their own particular time structure; they occupy scientific interest because of what they might become, but what they might become is unknown. Intentional action on the part of the scientist thus often ends in radical surprise at how the living thing exists after intervention. Rather than being utterly malleable and turning out as constructed or intended, things are resistant, resilient, and recalcitrant. See H. J. Rheinberger, "Cytoplasmic Particles: The Trajectory of a Scientific Object," in Lorraine Daston, ed., *Biographies of Scientific Objects* (Chicago: University of Chicago Press, 2000), pp. 270–294.

20. Eduard Uhlenhuth, "Changes in Pigment Epithelium Cells and Iris Pigment Cells of Rana Pipiens Induced by Changes in Environmental Conditions," *Journal of Experimental Medicine* 24 (1916): 690. See also Paul Rabinow, *Making PCR: A Story of Biotechnology* (Chicago: University of Chicago Press, 1996) for a discussion of the milieu of a biotechnology company in the 1980s.

21. Miroslav Holub, *Shedding Life: Disease, Politics, and Other Human Conditions,* trans. D. Young (Minneapolis: Milkweed Editions, 1997), p. 56.

22. Ludwig von Bertanlanffy, *Problems of Life: An Evaluation of Modern Biological Thought* (London: Watts, 1952), p. 134.

23. Susan Squier, *Liminal Lives: Imagining the Human at the Frontiers of Biomedicine* (Durham, NC: Duke University Press, 2005), Elizabeth A. Grosz, *Time Travels: Feminism, Nature, Power* (Durham, NC: Duke University Press, 2005).

24. Rose, "Life, Reason and History," 161.

25. Robert Pollack, "Hormones in Vitro," *Science* 208 (1980): 392–393.

26. Albert Ebeling, "Dr. Carrel's Immortal Chicken Heart: Present, Authentic Facts about This Oft-Falsified Scientific 'Celebrity,'" *Scientific American* (January 1942): 22–24.

27. Donna Haraway, *Modest_Witness@Second_Millennium.FemaleMan©_Meets _OncoMouse™: Feminism and Technoscience* (New York: Routledge, 1997), p. 82.

28. Philip J. Pauly, "Modernist Practice in American Biology," in Dorothy Ross, ed., *The Origins of American Social Science* (Cambridge: Cambridge University Press, 1991), pp. 272–289.

29. Georges Canguilhem, "The Living and Its Milieu," trans. John Savage, *Grey Room* 1(3) (2001[1952]): 7–31.

30. See Bowker and Starr's analysis in *Sorting Things Out* of classificatory technologies for an example of studying the infrastructure that orders information, objects, and people into categories rather than the individual instances of those facts, objects, or persons. Geoffrey Bowker and Susan Leigh Starr, *Sorting Things Out: Classification and Its Consequences* (Cambridge: MIT Press, 2000). Compare also Cambrosio and Keating's notion of the "biomedical platform" in research that looks at the practices which link different laboratory, clinical, and public spaces together and keep bodies of patients, samples derived from them, test results, and clinical entities functioning coherently across diverse disciplinary, technical, and architectural spaces. Alberto Cambrosio and Peter Keating, *Biomedical Platforms: Realigning the Normal and the Pathological in Late Twentieth-Century Medicine* (Cambridge: MIT Press, 2003).

31. Margaret R. Murray and Gertrude Kopech, *A Bibliography of the Research in Tissue Culture, 1884–1950* (New York: Academic Press, 1953).

32. Eugene Garfield, *Citation Indexing: Its Theory and Applications in Science, Technology and Humanities* (New York: John Wiley & Sons, 1979).

33. James Watson, *The Double Helix: A Personal Account of the Discovery of DNA* (New York: New American Library, 1969).

34. Michel Foucault, *The Order of Things: An Archaeology of the Human Sciences* (New York: Pantheon Books, 1971).

35. Paul Rabinow, *Anthropos Today: Reflections on Modern Equipment* (Princeton: Princeton University Press, 2003), p. 15.

36. Gillian Beer, *Darwin's Plots: Evolutionary Narrative in Darwin, George Eliot and Nineteenth Century Fiction* (Cambridge: Cambridge University Press, 2000 [1983]).

37. Rose, "Life, Reason and History," 161.

38. Robert Kohler, *Lords of the Fly: Drosophila Genetics and the Experimental Life* (Chicago: Chicago University Press, 1994).

39. Hans-Jörg Rheinberger, *Toward a History of Epistemic Things: Synthesizing Proteins in the Test Tube* (Stanford: Stanford University Press, 1997).

40. In somewhat different phrasing, Donna Haraway has also drawn attention to the intersection of animal existence and human intervention in the bodies of laboratory animals, calling the genetically engineered OncoMouse™ "simultaneously a metaphor, a technology, and a beast living its many-layered life as best it can." Haraway, *Modest_Witness,* 83. For further work on organisms such as the laboratory mouse, see Karen Rader, *Making Mice: Standardizing Animals for Biomedical Research, 1900–1955* (Princeton: Princeton University Press, 2004); for the Wistar rat, see Bonnie Clause, "The Wistar Rat as a Right Choice: Establishing Mammalian Standards and the Ideal of a Standardized Mammal," *Journal of the History of Biology* 26 (1993): 329–349; and for flat worms (Planaria), see Gregg Mitman and Anne Fausto-Sterling, "Whatever Happened to 'Planaria'? C. M. Child and the Physiology of Inheritance," in Adele E. Clarke and Joan H. Fujimura, eds., *Right Tools for the Job: At Work in 20th-Century Life Sciences* (Princeton: Princeton University Press, 1992). The comparison of cultured cells and organisms such as rats or mice is not exactly straightforward, particularly because one can be a subset of the other—cell lines established by using tissues from laboratory mice are a typical example of using one well-characterized animal technology to make another.

41. Cambrosio and Keating, *Biomedical Platforms,* 3.

1. Autonomy

1. Ross G. Harrison, "The Cultivation of Tissues in Extraneous Media As a Method of Morphogenetic Study," *Anatomical Record* 6 (1912): 181–193, quotation, p. 184.

2. Edmond Perrier, "Le Monde Vivant," *Feuilleton du Temps,* February 1, 1912.

3. "Paris Doctors Ask Proof of Carrel: Skeptics Declare His Experiments on Heart Tissue Too Marvelous to Credit," *New York Times,* June 23, 1912, p. C4.

4. H. Woollard, *Recent Advances in Anatomy* (London: J. & A. Churchill, 1927); Susan M. Billings, "Concepts of Nerve Fiber Development, 1839–1930," *Journal of*

the History of Biology 4 (1971): 275–305; Jane Maienschein, "Ross Harrison's Crucial Experiment As a Foundation for Modern American Experimental Embryology" (Ph.D. Dissertation, Indiana University, 1978).

5. E. N. Willmer, "Introduction," in E. N. Willmer, ed., Cells and Tissues in Culture: Methods, Biology and Physiology, vol. 1 (New York: Academic Press, 1965); Jan A. Witkowski, "Ross Harrison and the Experimental Analysis of Nerve Growth: The Origins of Tissue Culture," in T. J. Horder, J. A. Witkowski, and C. C. Wylie, eds., A History of Embryology (Cambridge: Cambridge University Press, 1986), pp. 149–177.

6. Jane Maienschein, "Experimental Biology in Transition: Harrison's Embryology, 1895–1910," Studies in History of Biology 6 (1983): 107–127; Jane Maienschein, Transforming Traditions in American Biology, 1880–1915 (Baltimore: Johns Hopkins University Press, 1991); Donna Haraway, Crystals, Fabrics, and Fields: Metaphors of Organicism in Twentieth-Century Developmental Biology (New Haven: Yale University Press, 1976); Jane M. Oppenheimer, "Ross Harrison's Contributions to Experimental Embryology," Bulletin of the History of Medicine 40 (1966): 525–543.

7. Readers concerned with Harrison's career and biography, and his other work in embryology, should seek out the literature cited in earlier endnotes, which details these matters.

8. Henry Harris, The Cells of the Body: A History of Somatic Cell Genetics (Cold Spring Harbor: Cold Spring Harbor Laboratory Press, 1995).

9. Nick Hopwood, "Producing Development: The Anatomy of Human Embryos and the Norms of Wilhelm His," Bulletin of the History of Medicine 74 (2000): 29–79.

10. Ross Harrison, "Observations on the Living Developing Nerve Fiber," Proceedings of the Society for Experimental Biology and Medicine 4 (1907): 140–143, quotation, pp. 140–141.

11. For a more detailed account of this controversy and its role and place in Harrison's overall career, see Haraway, Crystals, Fabrics, and Fields; and Maienschein, "Ross Harrison's Crucial Experiment."

12. The most in-depth account of the proponents of the different theories, the method and results they mobilized to argue for their point of view, and the variations within the theories themselves is to be found in Billings, "Concepts of Nerve Fiber Development."

13. Otto Breidbach, "The Controversy on Stain Technologies—An Experimental Reexamination of the Dispute on the Cellular Nature of the Nervous System Around 1900," History and Philosophy of the Life Sciences 18 (1996): 195–212.

14. S. Ramón y Cajal, *New Ideas on the Structure of the Nervous System in Man and Vertebrates*, trans. Neely Swanson and Larry W. Swanson (Cambridge: MIT Press, 1990 [1894]).

15. After decades of work to establish the credibility, utility, and necessity of histological techniques, arguments about the reliability of histological results tended to focus on the differences between fixatives, stains, and protocols, not on the difference between using living versus dead specimens. By 1905, histological technique was regarded by one writer as respectable enough to have reached an ascendant point in a history that could be divided into "three great periods": the Primitive Period of rudimentary technique before 1854, the Middle Period of 1854–1884 characterized by the "Supremacy of the Carmine Stain," and the Modern Period, or "Era of Elective Methods," the author's own time, in which it was axiomatic that "refinement of method and extension of knowledge go hand in hand" (Clarence B. Farrar, "The Growth of Histological Technique during the Nineteenth Century," *Review of Neurology and Psychiatry* 3 (1905): 501–515, 573–594, quotation, p. 501). This is a good example of an emphasis on refinement of the technique, to the exclusion of considerations of the technique's overall suitability, which was assumed. A history of the establishment of histology to this degree of acceptance and widespread use is outside of the scope of this book, but other historians have pointed to the amount of social and technical work it took to establish the merits and superiority of histology in the latter half of the nineteenth century. See L. Stephen Jacyna, "'A Host of Experienced Microscopists': The Establishment of Histology in Nineteenth-Century Edinburgh," *Bulletin of the History of Medicine* 75 (2001): 225–253; Bruce Bracegirdle, *A History of Microtechnique* (Ithaca: Cornell University Press, 1978); Nick Hopwood, "'Giving Body' to Embryos: Modeling, Mechanism and the Microtome in Late Nineteenth Century Anatomy," *Isis* 90 (1999):462–496.

16. P. Ehrlich and A. Lazarus, "Histology of the Blood: Normal and Pathological," in Fred Himmelweit, Martha Marquardt, and Henry Dale, eds., *Collected Papers of Paul Ehrlich*, vol. 1 (New York: Pergamon Press, 1956 [1900]), pp. 181–268, quotation, pp. 192–193.

17. Ehrlich and Lazarus, "Histology of the Blood," 193 (emphasis added).

18. S. Ramón y Cajal, *Recollections of My Life*, trans. E. Horne Craigie with Juan Cano (Cambridge: MIT Press, 1996 [1937]), quotation, pp. 526–527.

19. According to Susan Billings, "Cajal claimed that *Plasmodesmen* were artifacts of Held's preparative method, while Held, in turn, thought perhaps pale staining or alcohol fixation prevented Cajal from seeing them." Despite this, Cajal de-

clared, upon seeing Held's preparations, that "they show very much the same picture as ours" (Billings, "Concepts of Nerve Fiber Development," 296).

20. Ross Harrison, "The Outgrowth of the Nerve Fiber As a Mode of Protoplasmic Movement," *Journal of Experimental Zoology* 9 (1910): 787–846, quotation, p. 789.

21. Harrison, "The Outgrowth of the Nerve Fiber," 789.

22. Harrison, "Observations on the Living Developing Nerve Fiber," 141.

23. As his source for bacteriological technique, Harrison cites Robert Muir and James Ritchie, *Manual of Bacteriology,* 4th ed. (New York: MacMillan, 1907).

24. Harrison, "The Outgrowth of the Nerve Fiber," 801–802.

25. Harrison, "Observations on the Living Developing Nerve Fiber," 143.

26. "The Growth of Nerve Fibers," *Science* 30 (1909): 158 (emphasis added).

27. Harrison, "The Outgrowth of the Nerve Fiber," 840.

28. Harrison, "Observations on the Living Developing Nerve Fiber," 142.

29. Harrison, "The Outgrowth of the Nerve Fiber," 819.

30. John H. Hammond and Jill Austin, *The Camera Lucida in Art and Science* (Bristol, England: Adam Hilger, 1987).

31. Harrison, "The Outgrowth of the Nerve Fiber," 819.

32. Billings, "Concepts of Nerve Fiber Development," 296.

33. H. Braus, "Mikro-Kino-Projektionen von in vitro Gezüchteten Organanlagen," *Wiener Medizinische Wochenschrift* 61 (1911): 2809–2812, quotation, p. 2811. All translations, unless otherwise indicated, are my own.

34. Maienschein, "Experimental Biology in Transition," 119.

35. Ross G. Harrison, "Embryonic Transplantation and the Development of the Nervous System," in *The Harvey Lectures, Series 4* (New York: Academic Press, 1908), pp. 199–222.

36. William G. MacCallum to Ross Harrison, "circa 1909," Ross Harrison Papers, Sterling Memorial Library, Yale University, Accession #263, Record Unit III, Box 40, folder 361 (emphasis added).

37. Margaret R. Lewis and Warren H. Lewis, "The Growth of Embryonic Chick Tissues in Artificial Media, Agar, and Bouillon," *Johns Hopkins Hospital Bulletin* 22 (1910): 126–127.

38. Alexis Carrel and Montrose Burrows, "Cultivation of Tissues in Vitro and Its Technique," *Journal of Experimental Medicine* 13 (1911): 387–396, quotation, p. 388.

39. Ibid., p. 388.

40. Montrose Burrows to Ross Harrison, March 22, 1910. Ross Harrison Papers,

Sterling Memorial Library, Yale University, Accession #263, Record Unit III, Box 40, folder 346.

41. *American Men of Science* (New York: Bowker, 1921), p. 103.

42. Montrose Burrows, "The Cultivation of Tissues of the Chick-Embryo Outside the Body," *Journal of the American Medical Association* (December 10, 1910): 2057–2058, quotation, p. 2057.

43. Gerald L. Geison, *Michael Foster and the Cambridge School of Physiology: The Scientific Enterprise in Late Victorian Society* (Princeton: Princeton University Press, 1978).

44. Montrose Burrows, "The Tissue Culture As a Physiological Method," *Transactions of the Congress of American Physicians and Surgeons* 9 (1913):77–90, quotation p. 89.

45. Alexis Carrel and Montrose Burrows, "Cultivation of Adult Tissues and Organs Outside of the Body," *Journal of the American Medical Association* 55 (1910): 1379–1381.

46. Alexis Carrel and Montrose Burrows, "Cultivation of Sarcoma Outside of the Body: A Second Note," *Journal of the American Medical Association* 55 (1910): 1554.

47. Alexis Carrel and Montrose Burrows, "Artificial Stimulation and Inhibition of the Growth of Normal and Sarcomatous Tissues," *Journal of the American Medical Association* 56 (1911): 32–33, quotation, p. 33.

48. Carrel and Burrows, "Cultivation of Adult Tissues and Organs Outside of the Body," 1381.

49. Ibid., 1380.

50. Ibid., 1381.

51. Carrel and Burrows, "Human Sarcoma Cultivated Outside of the Body," 1732.

52. Ibid.

53. Carrel and Burrows, "Cultivation of Tissues in Vitro and Its Technique," 387.

54. Ibid., 388 (emphasis added).

55. Ross Harrison, "The Life of Tissues Outside the Organism from the Embryological Standpoint," *Transactions of the Congress of American Physicians and Surgeons* 9 (1913): 63–75, quotation, p. 65.

56. See Witkowski, "Ross Harrison and the Experimental Analysis of Nerve Growth"; Maienschein, "Experimental Biology in Transition"; Oppenheimer, "Ross Harrison's Contributions to Experimental Embryology." Harrison was particularly influenced by the work of Gustav Born in the practice of heteroplastic

grafting. According to Harrison, "heteroplastic grafting may be defined as the union of parts of organisms of different species into a single individual, or the combination of individuals of different species in double or multiple organisms living parabiotically" (Ross Harrison, "Heteroplastic Grafting in Embryology," in Sally Wilens, ed., *Organization and Development of the Embryo* (New Haven: Yale University Press, 1969 [1933]), p. 215). Born had accidentally observed in 1894 that severed parts of tadpoles would join up and heal together at the cut edges, even if the parts were from different species (G. Born, "Über Verwachsungsversuche mit Amphibienlarven," *Roux Archiv für Entwicklungsmechanik der Organismen* 4 (1897): 349–465, 517–623). By placing any two wounded edges together and keeping them in place for a time by binding the parts with silver wire, all manner of pieces of tissue could be grafted together. Others who took up the practice, such as Thomas Hunt Morgan and Harrison after Born's death in 1900, could, by grafting together parts of differently pigmented species, trace the fate of the cells of the respective species in the same body (Morgan, "Regeneration of Tissue Composed of Parts of Two Species," *Biological Bulletin* 1 (1899): 7–14).

57. Harrison was trained in medicine at Johns Hopkins Hospital, under the tutelage of Franklin Paine Mall, and he spent several years in the 1890s in Bonn working with Moritz Nussbaum on the development of the fins of Teleosts (J. S. Nicholas, "Ross Granville Harrison, January 13, 1870–September 30, 1959," *Biographical Memoirs, National Academy of Sciences USA* 35 (1961): 132–162). In Germany Harrison was strongly influenced by the work of Wilhelm Roux and the experimental methodology of the *Entwicklungsmechanik* approach more generally.

58. Fredrick Churchill, "Regeneration, 1885–1901," in Charles Dinsmore, ed., *A History of Regeneration Research: Milestones in the Evolution of a Science* (Cambridge: Cambridge University Press, 1991), pp. 113–132, quotation, p. 113.

59. Ross Harrison to William K. Brooks, April 23, 1896. The Alan Mason Chesney Medical Archives of The Johns Hopkins Medical Institutions.

60. Anne-Marie Moulin, *Le dernier langage de la médecine: histoire de l'immunologie de Pasteur au Sida* (Paris: Presses Universitaires de France, 1991). See Chapter 7.

61. Jane M. Oppenheimer, "Taking Things Apart and Putting Them Together Again," *Bulletin of the History of Medicine* 52 (1978): 149–161.

62. Jan Witkowski, "Experimental Pathology and the Origins of Tissue Culture: Leo Loeb's Contribution," *Medical History* 27 (1983): 269–288; Philip Rubin, "Leo Loeb's Role in the Development of Tissue Culture," *Clio Medica* 12 (1977): 33–56. Witkowski and Rubin differ in their opinion on the significance of Leo Loeb's work with explants and transplants to the history of tissue culture. Leo Loeb's brother, Jacques Loeb, claimed that "the method of cultivating tissue cells

in a test tube, in the same way as is done for bacteria, was first proposed and carried out by Leo Loeb, in 1897," and later "modified" by Harrison (Jacques Loeb, *The Organism as a Whole; From a Physico-chemical Viewpoint* (New York: Putnam, 1916), p. 31).

63. Georges Canguilhem, *Études d'histoire et de philosophie des sciences*, 3rd ed. (Paris: Librairie Philosophique J. Vrin, 1975); Albert Fischer, *Tissue Culture: Studies in Experimental Morphology and General Physiology of Tissue Cells in Vitro* (London: William Heinemann, 1925).

64. The French histologist Christian Champy, one of the first French scientists to take up tissue culture, wrote in 1913 that cell survival and organ transplantation were closely linked experimental phenomena: "Experiments of grafting are actually closely connected to those of cell survival. All grafts necessitate the survival of the tissue or of the organ grafted, not only during the time during which the organ is transported from one part to another of the body . . . but also during the time necessary for a new vascularization to be established . . . a grafted organ is not on first examination in very different conditions than those found in an organs submerged in plasma and placed in an incubator at 37 degrees." Christian Champy, "La survie et les cultures des tissus en dehors de l'organisme," *Le Mouvement Médical* (April 1913): 170–183, quotation p. 130.

65. Silas P. Beebe and James Ewing, "A Study of the Biology of Tumour Cells," *The British Medical Journal* (December 1, 1906): 1559–1560; Gottlieb Haberlandt, "Experiments on the Culture of Isolated Plant Cells," trans. A. D. Krikorian and D. L. Berquam, *The Botanical Review* 35 (1969 [1902]): 59–88; Ekkehard Höxtermann, "Cellular 'Elementary Organisms' in vitro. The Early Vision of Gottlieb Haberlandt and Its Realization," *Physiologia Plantarum* 100 (1997): 716–738.

66. As Paul Rabinow observes, writing of PCR and genome mapping in France, historians and anthropologists have not come up with very adequate tools for analyzing the event of the new in biology: "From time to time, new forms emerge that have something significant about them, something that catalyzes previously present actors, things, institutions into a new mode of existence, a new assemblage, an assemblage that made things work in a different manner. A manner that made many other things more or less suddenly possible. Such happenings are not reducible to the elements involved any more than they are representative of the epoch. Nor are such events mysterious and unanalyzable. It is only that so much effort has been devoted in the name of social science to explaining away the emergence of new forms as the result of something else that we lack adequate means

to conceptualize the event of new forms as the curious and potent singularity that it is" (Paul Rabinow, "Epochs, Presents, Events," in Margaret Lock, Allan Young, and Alberto Cambrosio, eds., *Living and Working with the New Medical Technologies: Intersections of Inquiry* (Cambridge: Cambridge University Press, 2000), pp. 31–46, quotation, p. 44).

67. "Growing Animal Tissues Outside of the Body," *Journal of the American Medical Association* 56 (1911): 1722–1723, quotation, p. 1722.

68. Albert Fischer, *Biology of Tissue Cells* (Copenhagen: Gylaeudalske Boghandel Norkisk Verlag, 1946), quotation, p. 327.

69. Canguilhem, *Études d'histoire et de philosophie des sciences,* 331.

70. Justin Jolly, "Sur la durée de la vie et de la multiplication des cellules animales en dehors de l'organisme," *Comptes rendus des Séances de la Société de Biologie* (November, 7, 1903): 1266–1267, quotation, p. 1267.

71. Justin Jolly, "À propos des communications de MM. Alexis Carrel et Montrose T. Burrows sur la 'Culture des Tissus,'" *Comptes rendus des Séances de la Société de Biologie* 69 (1910): 470–473, quotation, p. 473.

72. "Growing Animal Tissues Outside of the Body," *Journal of the American Medical Association,* quotation, p. 1722. Quoted in Witkowski, "Experimental Pathology and the Origins of Tissue Culture," 269.

73. Harrison, "The Life of Tissues Outside the Organism from the Embryological Standpoint," 63–64.

74. Oppenheimer, "Taking Things Apart and Putting Them Together Again."

75. There is a great deal of material written about Claude Bernard and his positing of the *milieu intérieur* as one of the bases of modern experimental medicine. See Frederic L. Holmes, "Claude Bernard and the Milieu Intérieur," *Archives Internationales d'Histoire des Sciences* 16 (1963): 369–376; Frederic L. Holmes, "Claude Bernard and Animal Chemistry," (Cambridge: Harvard University Press, 1974); Joseph Schiller, "Claude Bernard and the Cell," *The Physiologist* 4 (1961): 62–68; Alan G. Wasserstein, "Death and the Internal Milieu: Claude Bernard and the Origins of Experimental Medicine," *Perspectives in Biology and Medicine* 39 (1996): 313–326.

76. Claude Bernard, *Lectures on the Phenomena of Life Common to Animals and Plants,* trans. Hebbel E. Hoff, Roger Guillemin, Lucienne Guillemin (Springfield, IL: Charles C. Thomas, 1974 [1878]), p. 260.

77. Bernard, *Lectures on the Phenomena of Life,* 260.

78. François Jacob, *The Logic of Life: A History of Heredity,* trans. Betty E. Spillman (Princeton: Princeton University Press, 1973 [1970]), p. 188.

79. Harrison, "The Outgrowth of the Nerve Fiber," 799.

80. Maienschein, "Experimental Biology in Transition"; Oppenheimer, "Ross Harrison's Contributions to Experimental Embryology."

81. "Paris Doctors Ask Proof of Carrel," *New York Times*, C4.

2. Immortality

1. Lorraine Daston, ed., *Biographies of Scientific Objects* (Chicago: University of Chicago Press, 2000), p. 8.

2. Ibid., 6.

3. Philip J. Pauly, *Controlling Life: Jacques Loeb and the Engineering Ideal in Biology* (Berkeley: California University Press, 1990).

4. Alain Drouard, *Alexis Carrel (1873–1944): De la mémoire à l'histoire* (Paris: Éditions L'Harmattan, 1995); Susan Lederer, "Animal Parts/Human Bodies: Organic Transplantation in Early Twentieth-Century America," in Angela N. Creager and William C. Jordan, eds., *The Animal/Human Boundary: Historical Perspectives* (Rochester: University of Rochester Press, 2002), pp. 305–329; Theodor Malinin, *Surgery and Life: The Extraordinary Career of Alexis Carrel* (New York: Harcourt Brace, 1979); Shelley McKellar, "Innovation in Modern Surgery: Alexis Carrel and Blood Vessel Repair," in Darwin Stapleton, ed., *Creating a Tradition of Biomedical Research: Contributions to the History of the Rockefeller University* (New York: Rockefeller University Press, 2004), pp. 135–150.

5. Eduard Uhlenhuth, "Changes in Pigment Epithelium Cells and Iris Pigment Cells of Rana Pipiens Induced by Changes in Environmental Conditions," *Journal of Experimental Medicine* 24 (1916): 689–699.

6. Alexis Carrel, "Rejuvenation of Cultures of Tissues," *Journal of the American Medical Association* 57 (1911): 1611.

7. Alexis Carrel, "On the Permanent Life of Tissues Outside of the Organism," *Journal of Experimental Medicine* 15 (1912): 516–528, quotation, pp. 525–526.

8. Carrel, "On the Permanent Life of Tissues Outside of the Organism," 516.

9. Henry Field Smyth, "The Cultivation of Tissue Cells in Vitro and Its Practical Application," *Journal of the American Medical Association* 62 (1914): 1377–1381, quotation, p. 1377.

10. William Seifriz to Albert Ebeling, April 20, 1933. Alexis Carrel Papers, Lauinger Library, Georgetown University, Correspondence.

11. Alexis Carrel, "Contributions to the Study of the Mechanism of the Growth of Connective Tissue," *Journal of Experimental Medicine* 18 (1913): 289–300, quotation, p. 289.

12. Carrel, "Contributions to the Study," 300.

13. Alexis Carrel, "Physiological Time," *Science* 74 (1931):618–621.

14. Alexis Carrel, "Science Has Perfected the Art of Killing—Why Not of Saving?" *Surgery, Gynecology, and Obstetrics* 20 (1915): 710–711, quotation, p. 710.

15. Pierre LeComte du Noüy, *Biological Time* (New York: MacMillan, 1937), quotation, p. 69.

16. Carrel, "Physiological Time," 620.

17. Alexis Carrel, "Physiological Study of Tissues In Vitro," *Journal of Experimental Medicine* 38 (1923): 407–418, quotation, p. 408.

18. Alexis Carrel, "Tissue Culture and Cell Physiology," *Physiological Reviews* 4 (1924): 1–20, quotation, p. 1.

19. Jean Comandon, Constantin Levatidi, and Stefan Mutermilch, "Étude de la vie et de la croissance des dellules in vitro à l'aide de l'enregistrement cinématographique," *Comptes rendus des Séances de la Société de Biologie* 74 (1913): 464–467.

20. Carrel, "Physiological Time," 619.

21. Alexis Carrel, "The New Cytology," *Science* 73 (1931): 297–303, quotation, p. 303 (emphasis added).

22. Carrel, "The New Cytology," 298.

23. Carrel, "Physiological Time," 620.

24. That is, more specifically, in "The New Cytology," he argued that cells "are in physiological continuity with their environment. Cells and environment form a whole" [297]. That whole is the body, defined as a "system cells-environment" [298]. Cells condition their environment, and their environment affects cellular life in a mutual relation. The complexity of the "milieu intérieur" (and Carrel always specifically referred to Bernard's formulation of the body) conditioned by the different cells of the lungs, kidneys, liver, and intestines in the higher animal was made amenable to experimentation by tissue culture. Tissue culture maintained the "system cells-environment" and thus represented the processes of the body in undergoing life phenomena but was much simplified. An example of this is Carrel's use of tissue culture to study the phenomenon of "senescence": "The simplest artificial system which shows the phenomenon of aging consists of a colony of tissue or blood cells living in a medium limited in quantity. In such a system, the medium is progressively altered by the products of cell activity and, in its turn, reacts on the cells. Then aging and death take place" (Carrel, "Physiological Time," 621).

25. Carrel, "Physiological Time," 621.

26. Ibid.

27. Alexis Carrel, "Introduction," in Raymond C. Parker, *Methods of Tissue Culture* (New York: Paul B. Hoeber, 1938).

28. Alexis Carrel, "The Relation of Cells to One Another," in E. V. Cowdry, ed., *Human Biology and Racial Welfare* (New York: Paul B. Hoeber, 1930), pp. 205–218.

29. Carrel, "Physiological Time," 619.

30. Ross Harrison, "On the Status and Significance of Tissue Culture," *Archiv für Experimentelle Zellforschung* 6 (1927): 4–27, quotation, p. 18.

31. Jacques Loeb, *The Organism as a Whole; From a Physico-chemical Viewpoint* (New York: Putnam, 1916).

32. Raymond Pearl, *The Biology of Death* (Philadelphia: J. B. Lippincott, 1922).

33. For example, an anonymous review of J. B. S. Haldane's futuristic tract *Daedalus* in *Nature* in 1924 calmly declared that the predictions it made of growing humans in artificial wombs and generating the world's food supply by manipulating microorganisms was not far-fetched when one considered certain progress already made to date in biology. The reviewer asked his readers to think not of the artificial parthenogenesis of Loeb, or artificial contraception, or the transfer of live rabbit embryos from one rabbit's womb to another that Haldane himself mentions in the text, but of tissue culture: "We are introduced to artificially evolved Nitrogenfixing organisms, to the artificial alteration of character through applied endocrinology, and finally to 'ectogenesis'—the artificial development of human embryos outside the body, from ova taken from ovaries cultivated *in vitro* and artificially fertilised. Far fetched, this last prophecy? *Not so very, if what has already been done with tissue-culture is remembered.*" ("Review of Daedalus, or Science and the Future," *Nature* 113 (1924): 740 (emphasis added)). Tissue culture's manipulation of life here serves as a legitimation for the realism of predictions of manipulating society through intervening technically in biology and creating "artificial" forms of food, character, and reproduction.

34. "A Long Stride in Surgery," *Indianapolis News,* June 10, 1912. Alexis Carrel Papers, Lauinger Library, Georgetown University, Box 73.

35. Donna Haraway, *Modest_Witness@Second_Millennium.FemaleMan©_Meets_OncoMouse™: Feminism and Technoscience* (New York: Routledge, 1997), quotation, p. 82.

36. Of course, the two are not mutually exclusive. I have chosen this emphasis on engineering and invention because it makes more sense in the historical context already detailed by Philip Pauly, and it is the predominant narrative in the newspaper and magazine coverage of Carrel, while stories of *de novo* construction of humans from scratch appear much less frequently in these accounts. Jon Turney offers a complementary reading in placing Carrel's work in a genealogy of

Frankenstein narratives in *Frankenstein's Footsteps: Science, Genetics and Popular Culture* (New Haven: Yale University Press, 1998).

37. Stephen Kern, *The Culture of Time and Space, 1880–1918* (Cambridge: Harvard University Press, 1983).

38. Philip J. Pauly, "Modernist Practice in American Biology," in Dorothy Ross, ed., *The Origins of American Social Science* (Cambridge: Cambridge University Press, 1991), pp. 272–289.

39. Pauly, "Modernist Practice in American Biology," 286.

40. Rockefeller Archive Center, Clipping book RU450 RG432c.

41. W. Bruce Fye, *The Development of American Physiology* (Baltimore: Johns Hopkins University Press, 1987), p. 229.

42. *Portland Telegram*, Oregon, May 2, 1912. Alexis Carrel Papers, Lauinger Library, Georgetown University, Box 72.

43. George W. Corner, *A History of the Rockefeller Institute, 1901–1953* (New York: Rockefeller Institute Press, 1964), p. 158.

44. All the articles mentioned in this paragraph were found in Box 72 of the Alexis Carrel Papers, Lauinger Library, Georgetown University. Not all the clippings have retained the record of their exact date or source, but all of these were published on May 2, 1912.

45. "Surgeon Transplants Various Living Organs from One Animal to Another: How the Scientific World Is Amazed at Work of Dr. Alexis Carrel—Former Chicagoan Cuts Kidneys from One Cat and Puts Them in Another, Where They Function Properly—Transfers Arteries Kept in Cold Storage to Living Animals—Arteries from Leg of Man Transplanted in Dogs—Veins and Arteries Interchanged," *Sunday Magazine, Saint Louis Post-Dispatch*, June 9, 1912.

46. Ibid. Alexis Carrel Papers, Lauinger Library, Georgetown University, Box 73.

47. "Life and Death in Marvelous New Light," *New York Sun*, June 6, 1912. Alexis Carrel Papers, Lauinger Library, Georgetown University, Box 72.

48. Edmond Perrier, "Le Monde Vivant," *Feuilleton du Temps*, February, 1, 1912. Alexis Carrel Papers, Lauinger Library, Georgetown University, Box 72.

49. Harrison, "On the Status and Significance of Tissue Culture," 18.

50. Harrison, "On the Status and Significance of Tissue Culture," 18; de Noüy, *Biological Time*, 104.

51. "'Immortality' Is Achieved in Chicken Heart," *New York Herald Tribune*, November 22, 1925. Alexis Carrel Papers, Lauinger Library, Georgetown University, Box 74.

52. "Movie Reveals Living Cells of Tissue," *New York Evening Journal*, August 29, 1929. Alexis Carrel Papers, Lauinger Library, Georgetown University, Box 75.

53. "Living Tissue Cells Shown in Movie," *New York Times*, August 29, 1929. Alexis Carrel Papers, Lauinger Library, Georgetown University, Box 75.

54. Gladys Cameron, *Essentials of Tissue Culture Technique* (New York: Farrar and Rinehart, 1935), p. xv.

55. Albert Ebeling, "Dr. Carrel's Immortal Chicken Heart: Present, Authentic Facts about This Oft-Falsified Scientific 'Celebrity,'" *Scientific American* (January 1942): 22–24, quotation, p. 22.

56. Ibid., 22.

57. Ibid.

3. Mass Reproduction

1. Undated manuscript, "Tissue Cultures in the Study of Immunity: Retrospection and Anticipation," Presidential Address, American Association of Immunologists. John Enders Papers, Sterling Memorial Library, Yale University, Box 93, folder 92.

2. Angela Creager, *The Life of a Virus: Tobacco Mosaic Virus As an Experimental Model, 1930–1965* (Chicago: University of Chicago Press, 2002).

3. Alexis Carrel and Montrose Burrows, "Human Sarcoma Cultivated Outside of the Body: A Third Note," *Journal of the American Medical Association* 55 (1910): 1732.

4. Constantin Levatidi, "Symbiose entre le virus de la poliomyélite et les cellules des ganglions spinaux, à l'état de vie prolongée *in vitro,*" *Comptes rendus des Séances de la Société de Biologie* 74 (1913): 1179.

5. H. B. Maitland and M. C. Maitland, "Cultivation of Vaccinia Virus Without Tissue Culture," *Lancet* 2 (1928): 596.

6. Max Thieler and Hugh H. Smith, "The Use of Yellow Fever Virus Modified by in vitro Cultivation for Human Immunization," *Journal of Experimental Medicine* 65 (1937): 787–800; P. Mortimer, "Introduction and Commentary on 'The Use of Yellow Fever Virus Modified by in vitro Cultivation for Human Immunization,'" *Reviews in Medical Virology* 10 (2000): 3–16.

7. For a discussion of Peter Olitsky's claim to have cultured viruses in cell-free media, see Creager, *The Life of a Virus*, pp. 38–42.

8. George O. Gey, "An Improved Technic for Massive Tissue Culture," *American Journal of Cancer* 17 (1933): 752–756.

9. Ibid., 752.

10. Ibid., 754.

11. Ibid.

12. Ibid., 756.

13. Hannah Landecker, "The Lewis Films: Tissue Culture and 'Living Anatomy' at the Department of Embryology, 1919–1940," in Jane Maienschein, Marie Glitz, and Garland Allan, eds., *Centennial History of the Carnegie Institute, Volume 5: Department of Embryology* (Cambridge: Cambridge University Press, 2005), pp. 117–144.

14. Warren H. Lewis, "Pinocytosis," *Johns Hopkins Hospital Bulletin* 49 (1929): 17–27.

15. Lewis, "Pinocytosis," 18.

16. "Hydrophagocytosis," *Journal of the American Medical Association* 95 (1930): 1509.

17. Lewis, "Pinocytosis," 18.

18. Ibid.

19. Hans Zinsser and Emanuel Schoenbach, "Studies on the Physiological Conditions Prevailing in Tissue Cultures," *Journal of Experimental Medicine* 66 (1937): 207–227, quotation, p. 207.

20. Alexis Carrel, "Tissue Culture in the Study of Viruses," in Thomas Rivers, ed., *Filterable Viruses* (London: Balliere, Tindall & Cox, 1928), quotation, p. 104.

21. A. E. Feller, John Enders, and Thomas H. Weller, "The Prolonged Coexistence of Vaccinia Virus in High Titre and Living Cells in Roller Tube Cultures of Chick Embryonic Tissues," *Journal of Experimental Medicine* 72 (1940): 367–388, quotation, p. 384. It is clear from the drafts of papers in the Enders archive that Enders himself designed the experiments and wrote the papers with data generated by his junior coworkers. A biographical memoir for Enders authored by his former junior associates and fellow Nobel Prize recipients, Thomas H. Weller and Frederick C. Robbins, indicates that Enders did much more writing and directing of experiments than hands-on benchwork: "He would arrive in the middle of the morning carrying a simple lunch. His first priority was always to review any new observations. Although at the time he had no technician and only rarely participated actively in work at the bench, he delighted in looking at cultures and analyzing new data and knew exactly what was going on in his laboratory." Thomas H. Weller and Frederick C. Robbins, "John Franklin Enders: February 10, 1897–September 8, 1985," *Biographical Memoirs, National Academy of Sciences USA* 60 (1991): 46–65, quotation, p. 50.

22. Raymond C. Parker, "The Cultivation of Tissues for Prolonged Periods in Single Flasks," *Journal of Experimental Medicine* 64 (1936): 121–130.

23. Ibid., 124.

24. Feller, Enders, and Weller, "The Prolonged Coexistence of Vaccinia Virus," 368 (emphasis added).

25. Weller and Robbins, "John Franklin Enders," 50.

26. A wealth of books on polio and its significance in midcentury America were published in 2005, reflecting not just the fifty-year mark for the release of the polio vaccine but also a generation of authors who were children in this era and a renewed interest in vaccines and vaccine production. David Oshinsky, *Polio: An American Story* (New York: Oxford University Press, 2005); Marc Shell, *Polio and Its Aftermath: The Paralysis of Culture* (Cambridge: Harvard University Press, 2005); Paul Offit, *The Cutter Incident: How America's First Polio Vaccine Led to the Growing Vaccine Crisis* (New Haven: Yale University Press, 2005). For a more popular account that repeats the focus on Jonas Salk and the polio vaccine, see Jeffrey Kluger, *Splendid Solution: Jonas Salk and the Conquest of Polio* (New York: Putnam, 2005).

27. John Enders, Fredrick C. Robbins, and Thomas H. Weller, "The Cultivation of the Poliomyelitis Viruses in Tissue Culture," in *Nobel Lectures Physiology or Medicine 1942–1962* (Amsterdam: Elsevier, 1964) pp. 448–467.

28. Ibid., 453.

29. Harvey Blank, Lewis Corriell, and T. F. McNair Scott, "Human Skin Grafted Upon the Chorioallantois of the Chick Embryo for Virus Cultivation," *Proceedings of the Society for Experimental Biology and Medicine* 69 (1948): 341–345.

30. Ernest Goodpasture, Beverly Douglas, and Katherine Anderson, "A Study of Human Skin Grafted upon the Chorio-Allantois of Chick Embryos," *Journal of Experimental Medicine* 68 (1938): 891–904.

31. Ibid., 891–892.

32. For a description of the initial tests Jonas Salk did with the polio vaccine on institutionalized children, see Jane Smith, *Patenting the Sun: Polio and the Salk Vaccine* (New York: William Morrow, 1990). For a more general discussion of human experimentation, see Susan Lederer, *Subjected to Science: Human Experimentation in America Before the Second World War* (Baltimore: Johns Hopkins University Press, 1995).

33. James Jones, *Bad Blood: The Tuskegee Syphilis Experiment*, 2nd ed. (New York: Free Press, 1993). Elizabeth Fee, in an analysis for the city of Baltimore in the first half of the twentieth century, writes that the situation there was similar to that in the South: Venereal diseases among the black population were seen "as both evidence and consequence of their promiscuity, sexual indulgence, and immorality." Elizabeth Fee, "Venereal Disease: The Wages of Sin?" in Kathy Peiss and Christina Simmons, eds., *Passion and Power: Sexuality in History* (Philadelphia: Temple Uni-

versity Press, 1989), p. 182. In the context of this history, Lacks's being sent to the syphilis clinic reveals to some extent the perception of the patient by the doctor; furthermore, when this event reappears later as part of the narrative of the origin of the HeLa cell, the cell line is personified with metaphors of promiscuity and contagion (see Chapter 4).

34. William Scherer, Jerome Syverton, and George Gey, "Propagation *in vitro* of Poliomyelitis Viruses in Cultures of Human Epithelial Cells (Strain HeLa, Gey) Derived from Carcinoma of the Cervix," *Federal Proceedings* 12 (1953): 457.

35. Jerome Syverton and William Scherer, "Utilization of a Stable Strain of Human Epithelial Cells (HeLa, Gey) for Diagnosis of Poliomyelitis," *Federal Proceedings* 12 (1953): 462.

36. Ibid.

37. When this method was taken up in French virus research, it was changed to suit local conditions; children's tonsils were used instead of fetal tissues, which were hard to get in France. Jean-Paul Gaudillière, "Paris-New York Roundtrip: Transatlantic Crossings and the Reconstruction of the Biological Sciences in Post-War France," *Studies in History and Philosophy of Biological and Biomedical Sciences* 33 (2002): 389–417. Gaudillière describes the method as the "Enders rolling tube" technique, which shows the international visibility Enders's work gave to the tissue culture work of less prominent scientists such as George Gey; quotation, p. 401.

38. Thomas H. Weller, "The Application of Tissue-Culture Methods to the Study of Poliomyelitis," *New England Journal of Medicine* 249 (1953): 186–195.

39. Smith, *Patenting the Sun*, p. 126.

40. John Paul, *A History of Poliomyelitis* (New Haven: Yale University Press, 1977).

41. Jean-Paul Gaudillière and Ilana Löwy, eds., *The Invisible Industrialist: Manufactures and the Production of Scientific Knowledge* (New York: St. Martin's Press, 1998).

42. Enders to Feller 9.12.40. John Enders Papers, Sterling Memorial Library, Yale University, Box 24, folder 568.

43. John Enders to Microbial Associates 2.19,54. John Enders Papers, Sterling Memorial Library, Yale University, Box 52, folder 1259.

44. See, for example, Albert Fischer (a disciple of Carrel) writing in *Biology of Tissue Cells* in 1946: "Although the cells possess considerable mobility, they nevertheless remain part of a tissue; the migration away from the mother fragment is not too great to prevent them from maintaining their mutual cytoplasmic connections. This circumstance must undoubtedly be considered an expression for strong

positive affinities between cells, making possible a cooperation without which they would soon perish. Often enough it has been observed that isolated cell individuals are unable to form new communities." Albert Fischer, *Biology of Tissue Cells* (Copenhagen: Gyldendalske Boghandel Norkisk Forlag, 1946), quotation, p. 331.

45. For a history of Keith Porter and the electron microscope, see Nicholas Rasmussen, *Picture Control: The Electron Microscope and the Transformation of Biology in America, 1940–1960* (Stanford: Stanford University Press, 1997).

46. Quoted in Carol L. Moberg, "Keith Porter and the Founding of the Tissue Culture Association: A Fiftieth Anniversary Tribute, 1946–1996," *In Vitro Cell and Developmental Biology* 32 (1996): 663–669, quotation, p. 665.

47. Quoted in Moberg, "Keith Porter and the Founding of the Tissue Culture Association," 666.

48. Carol Moberg, who has detailed Keith Porter's role in the founding of the Tissue Culture Association, argues that once the service role of the organization had been quickly and successfully addressed, the organization served as the basis for the disciplinary formation of "a new science of cell biology." Moberg, "Keith Porter and the Founding of the Tissue Culture Association."

49. William F. Scherer, ed., *An Introduction to Cell and Tissue Culture* (Minneapolis: Burgess Publishing, 1955).

50. Margaret R. Murray and Gertrude Kopech, *A Bibliography of the Research in Tissue Culture, 1884–1950* (New York: Academic Press, 1953).

51. Daniel Kevles and Gerald Geison, "The Experimental Life Sciences in the Twentieth Century," *Osiris* 10 (1995): 97–121, quotation, p. 111.

52. Russell Brown and James Henderson, "The Mass Production and Distribution of HeLa Cells at Tuskegee Institute, 1953–55," *Journal of the History of Medicine* 38 (1986): 415–431.

53. Jacques Kelly, "Her Cells Made Her Immortal," *Baltimore Sun*, March, 18, 1997, 1A.

4. HeLa

1. Karen Rader, *Making Mice: Standardizing Animals for Biomedical Research 1900–1955* (Princeton: Princeton University Press, 2004); Robert Kohler, *Lords of the Fly: Drosophila Genetics and the Experimental Life* (Chicago: Chicago University Press, 1994); Robert Kohler, "Systems of Production: Drosophila, Neurospora and Biochemical Genetics," *Historical Studies in the Physical and Biological Sciences* 22 (1991): 87–130.

2. Albert Fischer, who studied tissue culture methods with Alexis Carrel, wrote that the issue was "whether the tissue culture is to be regarded as an organism-like system, or as a colony of independent cell elements," in his case coming down on the side of the organism-like system. Albert Fischer, *Biology of Tissue Cells* (Copenhagen: Gyldendalske Boghandel Norkisk Forlag, 1946), quotation, p. 331. See also the discussion of the various hypotheses regarding single-cell culture in Katherine Sanford, Wilton Earle, and Gwendolyn Likely, "The Growth *in vitro* of Single Isolated Tissue Cells," *Journal of the National Cancer Institute* 9 (1948): 229–246.

3. Sanford, Earle, and Likely, "The Growth *in vitro* of Single Isolated Tissue Cells."

4. Gordon Sato, "More Questions Than Answers," *In Vitro Cellular and Developmental Biology—Animal* 38 (2002): 429–435, quotation, p. 430.

5. Theodore T. Puck, *The Mammalian Cell as Microorganism: Genetic and Biochemical Studies in vitro* (San Francisco: Holden Day, 1972).

6. Richard G. Ham and William L. McKeehan, "Nutritional Requirements for Clonal Growth of Nontransformed Cells," in Hajim Katsuka, ed., *Nutritional Requirements of Cultured Cells* (Tokyo: Japan Scientific Societies Press, 1978) pp. 63–115, quotation, p. 64 (emphasis added).

7. Theodore T. Puck and Philip I. Marcus, "A Rapid Method for Viable Cell Titration and Clone Production with HeLa Cells in Tissue Culture, the Use of Irradiated Cells to Supply Conditioning Factors," *Proceedings of the National Academy of Sciences USA* 41 (1955): 432–437.

8. Harry Eagle, "Nutrition Needs of Mammalian Cells in Tissue Culture," *Science* 122 (1955): 501–504.

9. Eagle, "Nutrition Needs of Mammalian Cells in Tissue Culture," 501.

10. Indeed, this medium proved so easy to use and was so successful for culturing many kinds of cells that Richard Ham in 1978 commented that it may have inadvertently hindered further research into the subject: "the important contribution that Dr. Eagle made was to devise a very simple and reliable medium that, with some serum added, made it possible for many different people to use cell culture as a tool in many types of biological assay. He did that particular job so well that I sometimes feel he set back studies in cell nutrition. Cells would grow so well with some serum in his medium that many people lost interest in studies of cell nutrition, and it's only recently that a new interest has grown up." Ham and McKeehan, "Nutritional Requirements for Clonal Growth of Nontransformed Cells," 38.

11. Theodore Puck and Harold Fisher, "Genetics of Somatic Mammalian Cells

1. Demonstration of the Existence of Mutants with Different Growth Requirements in a Human Cancer Cell Strain (HeLa)," *Journal of Experimental Medicine* 104 (1956): 427–434.

12. Gordon Sato, Harold Fisher, and Theodore Puck, "Molecular Growth Requirements of Single Mammalian Cells," *Science* 126 (1957): 961–964, quotation, p. 962.

13. Ham and McKeehan, "Nutritional Requirements for Clonal Growth of Nontransformed Cells," 65.

14. Ann Kiessling and Scott Anderson, *Human Embryonic Stem Cells* (Sudbury, MA: Jones and Bartlett, 2003).

15. Georges Canguilhem, "The Living and Its Milieu," trans. John Savage, *Grey Room* 1(3) (2001[1952]): 7–31.

16. A. S. Parkes, "Cryobiology," *Cryobiology* 1 (1964): 18–39, quotation, p. 25.

17. J. K. Sherman, "Low Temperature Research on Spermatazoa and Eggs," *Cryobiology* 1 (1964): 103–129, quotation, p. 105.

18. Richard Billingham, "Concerning the Origins and Prospects of Cryobiology and Tissue Banks," *Transplantation Proceedings* 8(2) Suppl 1 (June 1976): 7–13, quotation, p. 7.

19. William Scherer and Alicia Hoogasian, "Preservation at Subzero Temperatures of Strain L and HeLa," *Proceedings of the Society for Experimental Biology and Medicine* 87 (1954): 480–487.

20. Scherer and Hoogasian, "Preservation at Subzero Temperatures," 483.

21. Richard Billingham and Peter Medawar, "The Freezing, Drying and Storage of Mammalian Skin," *Journal of Experimental Biology* 29 (1952): 454–468, quotation, p. 466.

22. Cell Culture Committee, "Animal Cell Strains," *Science* 146 (1964): 241–243.

23. Ibid., 241.

24. Ibid.

25. Ira Kline and Robert Trapani, "The Freezing and Preservation of Animal Tumors," *Cryobiology* 1 (1964): 130–139, quotation, p. 130.

26. Ward Peterson and Cyril Stulberg, "Freeze Preservation of Cultured Animal Cells," *Cryobiology* 1 (1964): 80–86, quotation, p. 85.

27. A. S. Parkes, quoted in Audrey U. Smith, *Biological Effects of Freezing and Supercooling* (Baltimore: Williams and Wilkins, 1961), p. 38.

28. Jennie P. Mather and Penelope E. Roberts, *Introduction to Cell and Tissue Culture: Theory and Technique* (New York: Plenum Press, 1998), p. 80.

29. Ibid., 85.

30. J. K. Sherman, "Low Temperature Research on Spermatazoa and Eggs," 103.

31. Alan Sterling Parkes, "Preservation of Spermatazoa, Red Blood Cells and Endocrine Tissue at Low Temperatures," in Robert Harris, ed., *Freezing and Drying* (New York: Hafner Publishing, 1952), pp. 99–106, quotation, p. 103.

32. Robert Pollack, *Readings in Mammalian Cell Culture* (Cold Spring Harbor: Cold Spring Harbor Laboratory Press, 1973), p. 3.

33. Frederick Bang, "History of Tissue Culture at Johns Hopkins," *Bulletin of the History of Medicine* 51 (1977): 516–537, quotation, p. 534. Trypsinizing means shaking the cells apart by using the digestive enzyme trypsin.

34. Charles Pomerat to George Gey, March 5, 1954. All letters cited in this chapter are from the George O. Gey Collection, The Alan Mason Chesney Medical Archives of The Johns Hopkins Medical Institutions.

35. Ronald H. Berg to George Gey, November 24, 1953. Berg was referring to a story that had appeared in the *Minneapolis Star* in 1954 that gave Lacks's name, although it is unclear who released the name. Gey's colleagues in Minneapolis denied revealing the name.

36. Bill Davidson, "Probing the Secret of Life," *Colliers* (May 14, 1954): 78–83, quotation, p. 79. After this point, there is a proliferation of pseudonyms made from the HeLa letters: Helen Lane and Helen Larson are two of these, probably spread by Gey himself, who was startled to find that Lacks's real name had leaked out without his knowledge. He thought that a fictitious name would serve equally well.

37. Davidson, "Probing the Secret of Life," 80.

38. John R. Masters, "HeLa Cells 50 Years On: the Good, the Bad, and the Ugly," *Nature Reviews Cancer* 2 (2001): 315–319.

39. Miroslav Holub, "Tissue Culture, or About the Last Cell," *New England Review/Bread Loaf Quarterly* 12 (1990): 376–379, quotation, p. 376.

40. Leonard Hayflick, "Biology of Human Aging," *American Journal of the Medical Sciences* 265 (1973): 433–445.

41. Henry Harris, *The Cells of the Body: A History of Somatic Cell Genetics* (Cold Spring Harbor: Cold Spring Harbor Laboratory Press, 1995), p. 45.

42. "Transformation" is the technical term used to describe a cellular event such as a mutation, chromosomal rearrangement, or viral infection, after which cells in culture grow to higher densities, in several layers rather than a monolayer, and cause tumors when injected into animals. This "event" can either occur spontaneously or be induced.

43. Stanley Gartler, "Genetic Markers as Tracers in Cell Culture," Second Decennial Review Conference on Cell Tissue and Organ Culture, *National Cancer Institute Monograph* 26 (1967): 167–195.

44. Ibid., 173.

45. Hannah Landecker, "Immortality, *in vitro:* A History of the HeLa Cell Line," in Paul Brodwin, ed., *Biotechnology and Culture: Bodies, Anxieties, Ethics* (Bloomington: Indiana University Press, 2000), pp. 53–72.

46. Stanley Gartler to George Gey, March 16, 1966. Emphasis added.

47. Barbara J. Culliton, "HeLa Cells: Contaminating Cultures Around the World," *Science* (June 7, 1974): 1059.

48. Howard W. Jones, Jr., Victor A. McKusick, Peter S. Harper, and Kuang-Dong Wuu, "After Office Hours: The HeLa Cell and a Reappraisal of Its Origin," *Obstetrics and Gynecology* 38 (1971): 945–949, quotation, p. 947.

49. BBC Radio interview with Adam Curtis (director of 1997 documentary about HeLa, "The Way of All Flesh"), April 14, 1997.

50. Walter Nelson-Rees, V. M. Zhdanov, P. Hawthorne, and Robert Flandermeyer, "HeLa-Like Marker Chromosomes and Type-A Variant Glucose-6-Phosphate Dehydrogenase Isoenzyme in Human Cell Cultures Producing Mason-Pfizer Monkey Virus-like Particles," *Journal of the National Cancer Institute* 53(3) (1974): 751–757.

51. Michael Rogers, "The HeLa Strain," *Detroit Free Press,* March 21, 1976, pp. 1D–4D, quotation, p. 4D, col. 5. The same article appeared under the title "The Double-Edged Helix" in the March 1976 issue of *Rolling Stone.*

52. Ibid., 1D.

53. Quoted in Michael Gold, *A Conspiracy of Cells: One Woman's Immortal Legacy and the Medical Scandal It Caused* (Albany: State University of New York Press, 1985) p. 63.

54. Ibid., 64.

55. Ibid., 72.

56. Masters, "HeLa Cells 50 Years On," 316.

57. Rob Stepney, "Immortal, Divisible: Henrietta Lacks Died 40 Years Ago But Her Cells Live On and Multiply," *The Independent,* March 13, 1994.

58. Charles Pomerat to George Gey, March 5, 1954.

59. Rebecca Skloot, "Henrietta's Dance," *Johns Hopkins Magazine* April (2000): 16–19.

60. Jones et al., "After Office Hours: The HeLa Cell and a Reappraisal of Its Origin," 945.

61. As was the case of the biopsy cells, this photograph is used in this and many other publications, such as a 1973 textbook of medical genetics, without any indication that permission was sought or given for its use, either from Lacks or her family.

62. Victor A. McKusick and Robert Claiborne, *Medical Genetics* (New York: HP Publishing, 1973), p. xv.

63. Laurence Karp, "The Immortality of a Cancer Victim Dead since 1951," *Smithsonian Magazine* March (1976): 51–56, quotation, p. 56.

64. Susan H. Hsu, Bernice Z. Schacter, Nancy L. Delaney, Thomas B. Miller, Victor A. McCusick, R. H. Kennet, J. G. Bodmer, D. Young, and W. F. Bodmer, "Genetic Characteristics of the HeLa Cell," *Science* 191 (1976): 392–394, quotation, p. 392.

65. Karp, "The Immortality of a Cancer Victim Dead since 1951," 55.

66. Skloot, "Henrietta's Dance," 19.

67. Ann Enright, "What's Left of Henrietta Lacks?" *London Review of Books* April 13, 2000, 8–10.

68. Masters, "HeLa Cells 50 Years On," 318. See also John Masters, "Human Cancer Cell Lines: Fact and Fantasy," *Nature Reviews: Molecular Cell Biology* 1(2000): 233–236.

5. Hybridity

1. Jan Sapp, *Beyond the Gene: Cytoplasmic Inheritance and the Struggle for Authority in Genetics* (Oxford: Oxford University Press, 1987); Evelyn Fox Keller, *Refiguring Life: Metaphors of Twentieth Century Biology* (New York: Columbia University Press, 1995).

2. The literature on model organisms is extensive and deals for the most part with the various organisms that have been selected as simple systems for the investigation of phenomena also found in more complex and experimentally cumbersome animals and humans. On the fruit fly, see Robert Kohler, *Lords of the Fly: Drosophila Genetics and the Experimental Life* (Chicago: Chicago University Press, 1994); on fungi, Robert Kohler, "Systems of Production: Drosophila, Neurospora and Biochemical Genetics," *Historical Studies in the Physical and Biological Sciences* 22 (1991): 87–130; on *C. elegans*, Soraya de Chadarevian, "Of Worms and Programmes: *Caenorhabditis elegans* and the Study of Developments," *Studies in the History and Philosophy of the Biological and Biomedical Sciences* 29 (1998):81–105; on mice, Karen Rader, *Making Mice: Standardizing Animals for Biomedical Research 1900– 1955* (Princeton: Princeton University Press, 2004).

3. Nick Hopwood, "Review of *The Cells of the Body*," *The Lancet* 347 (1996): 1317–1318.

4. Joseph Lucas and Naohiro Terada, "Cell Fusion and Plasticity," *Cytotechnology* 41 (2003):103–109. Lucas and Terada review the early years of cell fusion papers in relation to questions of cell plasticity in stem cell biology, because in their view "early results obtained by nuclear transplantation with amphibians combined with somatic cell fusion experiments provided evidence for the po-

tential pluripotency of somatic cell nuclei and for their capacity to be re-programmed." As such, they argue, these experiments "provided a background of both information and inspiration" for subsequent work; quotation, p. 107.

5. Lewis Thomas, *The Lives of a Cell: Notes of a Biology Watcher* (New York: Viking, 1974).

6. Anyone interested in the who-did-what-when narrative of somatic cell genetics in terms of key experiments and the sequence of events and findings in various places should consult the papers collected in Richard L. Davidson, *Somatic Cell Genetics* (Stroudsberg, PA: Hutchinson Ross, 1984); and the narrative of the field reported by one of its leading participants in Henry Harris, *The Cells of the Body: A History of Somatic Cell Genetics* (Plainview, NY: Cold Spring Harbor Laboratory Press, 1995). The role of Boris Ephrussi is detailed in Doris T. Zallen and Richard M. Burian, "On the Beginnings of Somatic Cell Hybridization: Boris Ephrussi and Chromosome Transplantation," *Genetics* 132 (1992): 1–8.

7. Jack Schultz, "Malignancy and the Genetics of the Somatic Cell," *Annals of the New York Academy of Sciences* 71 (1958): 994–1008.

8. J. B. S. Haldane, "Some Alternatives to Sex," *The New Biology* 19 (1955): 7–26.

9. Schultz, "Malignancy and the Genetics of the Somatic Cell," 995.

10. Guido Pontecorvo, *Trends in Genetic Analysis* (New York: Columbia University Press, 1958), quotation, p. 116.

11. Pontecorvo, *Trends in Genetic Analysis,* 116.

12. Guido Pontecorvo, "Methods of Microbial Genetics in an Approach to Human Genetics," *British Medical Bulletin* 18 (1962): 81–84, quotation, p. 81.

13. Joshua Lederberg, "Genetic Approaches to Somatic Cell Variation: Summary Comment," *Journal of Cellular and Comparative Physiology* 52 Supplement 1 (1958): 383–401, quotation, p. 384.

14. Haldane, "Some Alternatives to Sex," 21.

15. Theodore Puck, *The Mammalian Cell As Microorganism: Genetic and Biochemical Studies in vitro* (San Francisco: Holden Day, 1972). An auxotrophic mutant is a cell that has at least one more nutritional requirement to stay alive in culture than the cell type from which it was derived, presumably because a genetic mutation has occurred, making it different from its originating cell line. Bacterial auxotrophic mutants were an important part of the system for telling genetically different bacteria apart by an easily measured phenotypic difference: If the bacterium could grow only on nutrient medium supplied with that one required substance, it carried the mutation.

16. It was also the era of a search for tumor viruses, and the cellular techniques used by animal virologists such as George Gey, John Enders, and Renato Dulbecco

were part of looking for tumor viruses in cells. See Jean-Paul Gaudillière, "The Molecularization of Cancer Etiology in the Postwar United States: Instruments, Politics and Management," in Soraya de Chadarevian and Harmke Kamminga, eds., *Molecularizing Biology and Medicine: New Practices and Alliances 1910s–1970s* (Amsterdam: Harwood Academic Publishers, 1998), pp. 139–170; Angela Creager and Jean-Paul Gaudillière, "Experimental Arrangements and Technologies of Visualization: Cancer as a Viral Epidemic 1930–1960," in Jean-Paul Gaudillière and Ilana Löwy, eds., *Heredity and Infection: The History of Disease Transmission* (New York: Routledge, 2001), pp. 203–241.

17. Waclaw Szybalski and Elizabeth H. Szybalska, "Drug Sensitivity as a Genetic Marker for Human Cell Lines," in Donald Merchant and James V. Neel, eds., *Approaches to the Genetic Analysis of Mammalian Cells* (Ann Arbor: University of Michigan Press, 1962), pp. 11–27, quotation, p. 11.

18. This direct overlay of cellular phenotype as a transparent sign of genotype has been extensively critiqued and made much more complicated in recent years. For example, see Lenny Moss, *What Genes Can't Do* (Cambridge: MIT Press, 2002).

19. Georges Barski, Serge Sorieul, and Francine Cornefert, "Production dans des cultures in vitro de deux souches cellulaires en association, de cellules de caractère 'hybride,'" *Comptes rendus des séances de l'Académie des sciences* 251 (1960): 1825–1827.

20. Georges Barski, "Cytogenetic Alterations in Mixed Cultures of Mammalian Somatic Cells in vitro," in Robert Harris, ed., *Cytogenetics of Cells in Culture* (New York: Academic Press, 1964), pp. 1–11, quotation, p. 9.

21. Quoted in Richard M. Burian, Jean Gayon, and Doris Zallen, "Boris Ephrussi and the Synthesis of Genetics and Embryology," in Scott Gilbert, ed., *A Conceptual History of Embryology* (New York: Plenum Press, 1991), pp. 207–227, quotation, p. 215.

22. Boris Ephrussi, *Nucleo-Cytoplasmic Relations in Micro-Organisms* (Oxford: Clarendon Press, 1953), quotation, p. 108, quoted in Sapp, *Beyond the Gene*, 153.

23. Boris Ephrussi, *Croissance et régénération dans les cultures des tissus* (Paris: Gauthier-Villars & Cie, 1932).

24. Zallen and Burian, "On the Beginnings of Somatic Cell Hybridization," 1.

25. John Littlefield, "Selection of Hybrids from Matings of Fibroblasts in vitro and Their Presumed Recombinants," *Science* 145 (1964): 709–710.

26. Henry Harris and John F. Watkins, "Hybrid Cells Derived from Mouse and Man: Artificial Heterokaryons of Mammalian Cells from Different Species," *Nature* 205 (1965): 640–646, quotation, p. 646.

27. Boris Ephrussi and Mary C. Weiss, "Interspecific Hybridization of Somatic Cells," *Proceedings of the National Academy of Sciences (USA)* 53 (1965): 1040–1042,

quotation, p. 1040. Henry Harris claims in his autobiography that Ephrussi wrote him letters saying that Harris's results must have been in error, because a cross-species fusion was impossible. However, these letters are not currently available for examination.

28. George Yergenian and Mary Nell, "Hybridization of Dwarf Hamster Cells by UV-Inactivated Sendai Virus," *Proceedings of the National Academy of Sciences (USA)* 55 1966: 1066–1073, quotation, p. 1073.

29. Yergenian and Nell, "Hybridization of Dwarf Hamster Cells," 1066.

30. Boris Ephrussi and Mary C. Weiss, "Hybrid Somatic Cells," *Scientific American* 220 (1969): 26–35.

31. Roberta Bivins, "Sex Cells: Gender and the Language of Bacterial Genetics," *Journal of the History of Biology* 33 (2000):113–139.

32. Ephrussi and Weiss, "Hybrid Somatic Cells," 26.

33. "Cell Fusion: A New Gift to Biology," *Nature* 223 (1969): 1039–1041, quotation, p. 1039.

34. Davidson, *Somatic Cell Genetics,* 1.

35. Ephrussi and Weiss, "Hybrid Somatic Cells," 33.

36. Shapiro, "Final Discussion," in David Evered and Julie Whelan, eds., *Cell Fusion,* Ciba Foundation Symposium 103 (London: Pitman, 1984), quotation, p. 277.

37. Ephrussi and Weiss, "Hybrid Somatic Cells," 33 (emphasis added).

38. Henry Harris, *Nucleus and Cytoplasm* (Oxford: Clarendon Press, 1970), p. 110.

39. It is difficult to say, actually, whether the experiment was tried somewhere and failed without the extra help of the fusing agent. Experiments that do not work do not distinguish between inadequate technique and biological impossibility and are rarely published or otherwise publicized.

40. Henry Harris, *The Balance of Improbabilities: A Scientific Life* (Oxford: Oxford University Press, 1987), p. 186.

41. Harris and Watkins, "Hybrid Cells Derived from Mouse and Man," 640.

42. Mary C. Weiss and Boris Ephrussi, "Studies of Interspecific (Rat x Mouse) Somatic Hybrids. II. Lactate Dehydrogenase and ß-Glucuronidase," *Genetics* 54 (1966): 1111–1112.

43. Richard L. Davidson, Boris Ephrussi, and Kohtaro Yamamoto, "Regulation of Pigment Synthesis in Mammalian Cells, As Studied by Somatic Hybridization," *Proceedings of the National Academy of Sciences (USA)* 56 (1966): 1437–1440.

44. Vittorio Defendi, ed., *Heterospecific Genome Interaction* (Philadelphia: Wistar Institute Press, 1969).

45. Boris Ephrussi, *Hybridization of Somatic Cells* (Princeton: Princeton University Press, 1972), p. 30.

46. Ibid.

47. Harris, *Nucleus and Cytoplasm*, 112.

48. Lily Kay, *Who Wrote the Book of Life? A History of the Genetic Code* (Stanford: Stanford University Press, 2000).

49. Alberto Cambrosio and Peter Keating, *Exquisite Specificity: The Monoclonal Antibody Revolution* (Oxford: Oxford University Press, 1995); Cesar Milstein, "Monoclonal Antibodies," *Scientific American* 243 (1980): 56–64.

50. For a discussion of the long route monoclonal antibodies took from pure research to patented commodity, see Soraya de Chadarevian, *Designs for Life: Molecular Biology After World War II* (Cambridge: Cambridge University Press, 2002).

51. Indeed, in his Nobel Prize acceptance speech, Cesar Milstein emphasized the intellectual curiosity that drove these experiments in order to underscore the value of basic research, reflecting that hybridoma technology "resulted from esoteric speculations, for curiosity's sake; only motivated by a desire to understand nature." Quoted in de Chadarevian, *Designs for Life*, p. 359.

52. S. B. Carter, "Effects of Cytochalasins on Mammalian Cells," *Nature* 213 (1967) 261–264.

53. Thorfinn Ege and Nils Ringertz, "Preparation of Microcells by Enucleation of Micronucleate Cells," *Experimental Cell Research* 87 (1974):378–382, quotation, p. 381.

54. Clive L. Bunn, Douglas C. Wallace, and Jerome M. Eisenstadt, "Cytoplasmic Inheritance of Chloramphenicol Resistance in Mouse Tissue Culture Cells," *Proceedings of the National Academy of Sciences (USA)* 71 (1974):1681–1685.

55. Woodring E. Wright and Leonard Hayflick, "Nuclear Control of Cellular Aging Demonstrated by Hybridization," *Experimental Cell Research* 96 (1975): 113–121.

56. Bunn, Wallace, and Eisenstadt, "Cytoplasmic Inheritance of Chloramphenicol Resistance," 1681–1685.

57. Pontecorvo, *Trends in Genetic Analysis*, 116.

58. Adolf Graessmann and Monika Graessmann, "Microinjection Turns a Tissue Culture Cell into a Test Tube," in Jerry Shay, ed., *Techniques in Somatic Cell Genetics* (New York: Plenum Press, 1982), pp. 463–470, quotation, p. 463.

59. Hans-Jörg Rheinberger, "Beyond Nature and Culture: Modes of Reasoning in the Age of Molecular Biology and Medicine," in Margaret Lock, Allan Young, and Alberto Cambrosio, eds., *Living and Working with the New Medical Technologies* (Cambridge: Cambridge University Press, 2000), pp. 19–30.

60. There is also the question of what somatic cell hybridization had to do with a quite distinct but simultaneous rise in the use of the term "hybridization" to refer to the ability to stick single-stranded pieces of DNA or RNA to one another,

taking advantage of base complementarity. There is no explicit link between the two uses, though DNA and RNA hybridization techniques were used sometimes in cell hybridization experiments, particularly as a means to determine whether specific chromosomes had been transferred in a fusion or whether RNA was being synthesized in the hybrid cell.

61. Frank Ruddle, "Parasexual Approaches to Genetic Organization and Function: Introduction," in Roland F. Beers, Jr., and Edward G. Bassett, eds., *Cell Fusion: Gene Transfer and Transformation* (New York: Raven Press, 1984), pp. 79–80, quotation, p. 79.

62. Karl Illmensee, Peter Hoppe, and Carlo Croce, "Chimeric Mice Derived from Human-Mouse Hybrid Cells," *Proceedings of the National Academy of Sciences (USA)* 75 (1978): 1914–1918, quotation, p. 1917.

63. Karl Illmensee and Carlo Croce, "Xenogenic Gene Expression in Chimeric Mice Derived from Rat-Mouse Hybrid Cells," *Proceedings of the National Academy of Sciences (USA)* 76 (1978): 879–883.

64. Klaus Willecke, in discussion of Edward Cocking, "Plant-Animal Cell Fusions," in Evered and Whelan, eds., *Cell Fusion*, 127.

Epilogue

1. Rayna Rapp, *Testing Women, Testing the Fetus: The Social Impact of Amniocentesis in America* (New York: Routledge, 2000), quotation, p. 210.

2. Margaret Lock, "The Alienation of Body Tissue and the Biopolitics of Immortalized Cell Lines," in Nancy Scheper-Hughes and L. Wacquant, eds., *Commodifying Bodies* (London: Sage Publications, 2003), pp. 63–92, quotation, p. 71.

3. Paul Rabinow, *French DNA: Trouble in Purgatory* (Chicago: University of Chicago Press, 1999).

4. Paul Rabinow, *Essays on the Anthropology of Reason* (Princeton: Princeton University Press, 1996), quotation, p. 150.

5. Andrew Lakoff and Stephen Collier, "Ethics and the Anthropology of Modern Reason," *Anthropological Theory* 4 (2004): 419–434; Paul Rabinow, *Anthropos Today: Reflections on Modern Equipment* (Princeton: Princeton University Press, 2003).

6. Maria J. Martin, Alysson Muotri, Fred Gage, and Ajit Varki, "Human Embryonic Stem Cells Express an Immunogenic Nonhuman Sialic Acid," *Nature Medicine* 11 (2005): 228–232. This was recently discovered by scientists working with embryonic stem cells. The use of so-called feeder layers of irradiated, nonreplicating cells is part of growing embryonic stem cells, and these are most typically made using mouse cells. The media they are bathed in are also animal-derived.

Human embryonic stem cells grown using this "xenogenic culture methodology" would cause an immune reaction if used therapeutically in the human body, because humans do not make Neu5Gc and have antibodies specific to it. A human body would recognize them as foreign animal cells and reject them. The authors writing this report tested the commercial serum replacement media that they termed "standard" to human embryonic stem cell culture techniques. Because the techniques of standardized media and feeder layers emerged in the 1950s and were developed using animal cells, it is not surprising that the standardized tools that many scientists draw upon (or buy) in order to work with human cells are of animal origin and that the "human" cells produced thereby have adopted, to their cultivators' dismay, some of the characteristics of their animal culture environment. Only when a problem arises does this infrastructure become visible.

7. Sarah Franklin, "Review Essay: What We Know And What We Don't About Cloning and Society," www.comp.lancs.ac.uk/sociology/soc020sf.html (accessed 2001). A detailed and insightful discussion of cloning, agriculture, and the cultural significance of Dolly can be found in Sarah Franklin, *Dolly Mixtures: The Remaking of Genealogy* (Durham: Duke University Press, forthcoming 2007).

8. D. E. Pegg, "The History and Principles of Cryopreservation," *Seminars in Reproductive Medicine* 20 (2002): 5–14.

9. See the discussion of Peter Medawar's idea of the age chimera in Chapter 4.

10. I. L. Cameron and G. M. Padilla, *Cell Synchrony: Studies in Biosynthetic Regulation* (New York: Academic Press, 1966), quotation, p. vii.

11. Ian Wilmut, Keith Campbell, and Colin Tudge, *The Second Creation: Dolly and the Age of Biological Control* (Cambridge: Harvard University Press, 2000), quotation, p. 105.

12. Susan Squier, *Liminal Lives: Imagining the Human at the Frontiers of Biomedicine* (Durham: Duke University Press, 2004).

13. Gillian Beer, *Darwin's Plots: Evolutionary Narrative in Darwin, George Eliot and Nineteenth Century Fiction* (Cambridge: Cambridge University Press, 2000 [1983]), quotation, p. ix.

INDEX

alienation, 3, 19, 21, 69, 220, 221
American Type Culture Collection, 156, 179, 226
aneuploidy, 167
antibiotics, 103, 124, 131, 138, 155
artifact, 35, 132, 246n19
artifice, 16, 20, 93, 94, 104
artificial parthenogenesis, 5–6, 8, 94, 254n33
asepsis, 15, 38–40, 67, 83, 84, 103, 131
autonomy, 11, 15, 18, 61, 66, 69, 144, 147–148; of nerve or muscle cells in organs, 29–30, 49–50

bacteriology, 31, 39, 61, 78. *See also* microbiology
Barski, Georges, 190–193
Berg, Roland, 163–164
Bergson, Henri, 70, 82
Bergsonian philosophy, 71, 82
Bernard, Claude, 63, 78
Bichat, Xavier, 62
bioethics, 173, 222
biological modernism, 16, 92
biological supply industry, 132, 134, 156, 162
biological time, 11, 12, 71, 80, 85–87, 90, 154, 185, 227, 230–232
biotechnology, 1–2, 5; history of, 7, 23, 183,

221, 232; as milieu, 20, 126, 152, 242n20; and the human, 222–225, 232, 235
Blank, Harvey, 125
Braus, Hermann, 44, 46, 66
Burrows, Montrose, 48–55, 69

Cajal, Santiago Ramòn y, 35–37, 246n19
camera lucida, 42–43
Cameron, Gladys, 103
Campbell, Keith, 229
cancer: research, 54, 110, 126, 128, 178, 189, 201; cells, 57, 113–114, 127, 163, 167
Canguilhem, Georges, 6, 20, 24, 59, 152, 241n13
Carnegie Institution of Washington, 115
Carrel, Alexis, 15–16, 29, 68–71, 103–106, 118; establishment of tissue culture, 47–55, 72–79, 167; and control of biological time, 80–90, 99, 253n24; as public figure, 94–98, 102. *See also* immortal chicken heart; permanent life
Carrel flask, 82–84, 89, 105, 112–113
Carter, S. B., 211
cell: as technical object, 3, 13–14, 16, 26, 221; in history of biology, 4, 6–7, 11, 240n7; fusion, 19, 180–185, 192–216, 265n4; synchrony, 225, 228–229, 231–232. *See also* cancer; elementary organism; human cell culture; individuality, genetic

chimera, 154, 215, 203–204, 228
chinese hamster ovary (CHO) cell line, 152, 141
cinematography. *See* microcinematography
Claude, Albert, 133
cloning: somatic cell, 18, 143–147, 159–161, 177, 188–189; whole animal, 20–21, 224–225, 227–229, 232–233. *See also* individuality, genetic
contamination, 82, 154, 155, 158, 226; and HeLa, 166, 168–173, 178
Cornefort, Francine, 190–191
cryobiology, 159, 225, 231–232
culture medium, 11–12, 16, 27, 220; early experimentation with, 49, 73–79, 82–84, 113–117, 119–120, 253n24; synthetic, 79, 138; conditioned, 144–147, 150; defined, 148, 151–152, 189, 231, 261n10
cybrid, 206, 207, 212
cytochalasin b, 211

Dakin, Henry Drysdale, 80
Darwin, Charles, 24, 235
Davidson, Richard, 198
death, 32, 48, 62, 69; suppression of, 73–74, 76–77, 91; as technical failure, 92, 119; and HeLa, 142, 160–161, 164, 168, 176; and hybridity, 202, 217. *See also* life span
developmental biology, 7, 240n8
differentiation, 38, 54, 61, 73, 90, 180, 185, 191
disembodiment, 3–4, 14–15, 19, 142, 178, 213, 223, 225–226, 229
Dolly, 160, 225, 227, 271n7
Dulbecco, Renato, 192, 266n16
du Noüy, Pierre LeComte, 81, 100
duration, 13, 16, 70–71, 73–74, 87–90, 99, 226

Eagle, Harry, 148–149, 261n10
Earle, Wilton, 144, 147, 160
Ebeling, Albert, 76, 80, 104–105
Ehrlich, Paul, 35–36, 59
elementary organism, 6, 63, 194
embryo juice, 77, 79, 91, 148
embryology, 31, 48, 56, 61–62, 64, 115, 192

Enders, John: and tissue culture methods in virology, 17, 107, 112, 117–121, 131; and polio, 122–126, 129–130
engineering ideal, 10, 70
Ephrussi, Boris, 191–193, 195, 201–202, 205, 266n6
experimental system, 25, 186

Fauré-Fremiet, Emmanuel, 192
feeder layer, 150, 270n6
Flexner, Simon, 48
Foucault, Michel, 23–24, 241n13
Franklin, Sarah, 225, 271n7
freezing, 18, 142–143, 153–155, 157–160, 226–228
Frostie, 225, 227
fusogen, 197–198

Gartler, Stanley, 168–169
genetics, 6–8, 192; bacterial, 145, 186, 188–190, 196, 266n15; somatic cell, 181, 205, 212, 216, 266n6; without sex, 184, 186, 197–198, 217
Gey, George, 112–115, 117, 127–129, 138, 162–163, 173–174
Gey, Margaret, 18, 127
glucose-6 phosphate dehydrogenase (G6PD), 168–169, 171
Golgi, Camillo, 35
Goodpasture, Ernest, 125
Green, Howard, 197

Haldane, John Burdon Sanderson, 186, 188, 191, 254n33
Ham, Richard, 152, 261n10
Harris, Henry, 194–195, 199–201, 208, 212
Harrison, Ross: experiments with nerve culture, 14–16, 28–34, 37–46, 65–66, 80; in relation to Carrel, 47–50, 52–55, 70, 72; as founder of tissue culture, 50, 54, 133; reflections on significance of tissue culture, 55–58, 91, 100
Harvey Society, 46–47
Hayflick, Leonard, 166–167
Heartbeat, 49, 69, 76

HeLa, 18, 140; establishment and mass production, 127–129, 135–138; as standard
cell, 141, 147–149, 154, 156, 165; changing narratives of, 160–180; in cell fusion,
201, 208–209. *See also* Lacks, Henrietta
His, Wilhelm, 34
histology, 32, 35, 37–38, 41, 44, 66–67, 85–
86, 146n15
homology, 20, 182, 199, 224, 233
Hoogasian, Alicia, 154
human cell culture, 108, 124, 140, 143, 168,
186, 198, 223
hybrid: cell, 19–20, 181–182, 186, 192–193,
205–209, 217–218; vigor, 193; interspecies, 195, 202, 215. *See also* species
hybridity, 19, 183–184, 193, 202, 205–207,
216–218
hybridoma, 210, 269n51

immortal chicken heart, 16, 69, 90–93, 104,
160, 166–167, 176. *See also* public life of
biological entities
immortality, 68–73; as technical form of
life, 11, 16, 71, 80, 90, 103–104; of cultured cells, 19, 79, 89, 91, 167–168, 210;
in public discussion, 92–94, 99, 101; of
HeLa and Henrietta Lacks, 128, 164,
173–179. *See also* duration; permanent life
individuality, 29–30, 63, 104, 144, 200–201;
genetic, 18, 149–150, 177, 182, 187; of organisms, 19–20, 30, 57, 61, 138; human,
138, 142, 164, 174–175; cellular, 149–150,
177, 185, 190. *See also* cloning, somatic
cell
in vitro, 32, 50, 59–62, 78, 79, 87, 89, 90, 97,
213–214; shift from *in vivo*, 14–16, 33, 61,
66, 175, 215, 223

Jacob, François, 7, 63
Johns Hopkins University, 47–48, 112, 127–
128, 137, 140, 173–174, 193
Jolly, Justin, 59–62
Journal of Experimental Medicine, 96

Koch, Robert, 39, 95

L cell line, 141, 144–145, 148–149, 154
Lacks, Henrietta, 18, 127, 140–142, 168n69;
stories about, 163–164, 166, 170–178
Lederberg, Joshua, 88, 191
Levatidi, Constantin, 110
Lewis, Margaret Reed, 47, 115
Lewis, Warren, 47, 76, 115–116
life span: of cell versus body, 3, 11, 13–14,
19, 69; of cells in culture, 90–91, 108,
158, 177
Likely, Gwendolyn, 144, 160
Littlefield, John, 193
Loeb, Jacques, 1, 5, 10–11, 20, 70, 91, 94

MacCallum, William G., 46–47, 55, 58
Marcus, Philip, 145
mass reproduction, 137, 177
Masters, John, 178
Medawar, Peter, 154
Microbial Associates, 131, 134, 156, 162
microbiology, 135, 148, 193, 247n23. *See also*
bacteriology
microcell, 206, 212
microcinematography, 71, 84–88, 100, 115–
117, 133
microinjection, 213–214, 216
milieu, 20, 26, 103, 150, 152, 219, 223,
242n20
milieu intérieur, 32, 62, 65, 78, 151, 251n75,
253n24
model organism, 141, 181, 185, 265n2
molecular biology, 5, 6, 8, 143, 152, 183,
213, 216
monoclonal antibodies, 183–184, 210, 216,
269n50
Moore, John, 221
mumps virus, 121–122

National Cancer Institute, 144, 155–156, 170
National Foundation for Infantile Paralysis,
109, 122, 134–135, 163
National Institutes of Health, 135, 141, 148
National Research Council on Growth, 133
Nell, Mary, 195
Nelson-Rees, Walter, 170, 172–173
nerve development, 15, 28, 30–31, 33–46, 55

Nobel prize in medicine, 35, 71, 80, 94–95, 98, 123
nutrient medium. *See* culture medium
nutritional mutant, 149, 189, 266n15

O'Connor, Basil, 135
Oppel, Albert, 58
Oppenheimer, Jane, 57, 62

parabiosis, 65
parasexual approach, 185, 207
Parker, Raymond, 119–120, 133
passaging, 75, 110, 111, 155, 158, 230
Pasteur Institute (Paris), 95, 110
Pauly, Philip, 10, 16, 70, 94
Pearl, Raymond, 91
permanent life, 16, 70–72, 74–76, 95–97, 148, 166. *See also* immortality
personification, 142, 161, 164, 171–177, 179, 259n33
phagocytosis, 115
physiology, 78, 126, 148, 224
pinocytosis, 115–116
plasticity, 1, 8–13, 20–21, 27, 56–57, 104, 126, 161, 231–233; genealogy of, 8, 12, 20, 23; in cell fusion, 182, 194, 206, 217, 265n4
Polge, Christopher, 153, 226
polio, 17, 107–110; in cell culture, 122–125, 128–130; vaccine production, 135–139; relation to HeLa production, 162–165, 168, 176, 258n26
Pollack, Robert, 13, 160
Pomerat, Charles, 162, 190
Pontecorvo, Guido, 187–188, 191, 212
Porter, Keith, 132, 134
primary culture, 158, 165
protoplasmic bridge theory, 37–38, 44
public life of biological entities, 71, 92–93, 162, 164, 233
Puck, Theodore, 145, 149–150, 152, 189

Rabinow, Paul, 221, 250n66
race, 128, 169, 171–172

Rapp, Rayna, 219–220
recombinant DNA, 2, 182–183, 199, 214, 216
reconstituted cell, 20, 206–207, 218
recurrence, 6, 8, 24, 241n13
regeneration, 35, 47, 56–57, 71, 74, 76, 90, 192
rejuvenation, 69, 71, 73–74
reproductive science, 4, 7, 152, 227
Rheinberger, Hans-Jörg, 25, 214, 242n19
Robbins, Frederick, 121–122, 129, 257n21
Rockefeller Institute for Medical Research, 47–48, 70, 80, 88, 91, 94–95, 111, 125–126
roller tube culture method, 112–114, 117–119, 121, 124, 130, 259n37
Rous, Peyton, 51, 125
Rous sarcoma virus, 91, 167

Salk, Jonas, 130, 131, 135, 258n26
Sanford, Katharine, 144, 147, 160
Sapp, Jan, 191, 242n15
Scherer, William, 128–129, 153
Seifriz, William, 76
Smyth, Henry Field, 76
Sorieul, Serge, 190–191
species, 19–20, 57, 167, 182, 184–185, 194–197, 199–200, 205–206, 208; organization of research by, 222–224. *See also* hybrid, interspecies
stem cells, 4–5, 7, 151, 183, 232–233
surgery, 9, 47, 56, 71, 82–83, 95. *See also* transplantation
Syverton, Jerome, 128–129, 155

technology of living substance, 1, 20, 104
temporality, 1, 21, 27, 230, 232–233; of cells in culture, 11–13, 70–71, 85, 104; of experiment, 15, 31–32, 36, 41, 66, 72, 121, 201; of film, 46, 86; and freezing, 154, 157, 160–161, 177, 226, 228. *See also* biological time
Thomas, Lewis, 183
Tissue Culture Association, 134
totipotency, 6, 235
transgenesis, 182–184, 215–216, 225, 232

transplantation: in adult organisms, 9, 48, 56–57, 73–74, 200; in embryology, 44, 64–66, 192–193; tumor, 51, 64; nuclear, 216
Tuskegee Institute, 135–137, 147, 162

uncanny, 15, 69, 76, 97

vaccinia, 111, 119, 121, 123
varicella, 122
virology, 108, 120–123, 126, 134–135, 192, 201
vivisection, 78
von Bertalanffy, Ludwig, 13

Watkins, John, 194
Watson, John B., 94
Weiss, Mary, 195, 197, 201
Weller, Thomas, 121–122, 129–131
Wells, Herbert George, 9–11, 20–21
Wilmut, Ian, 229–230
World War I, 80, 82, 86
wound healing, 48, 71, 73, 89

yellow fever vaccine, 111
Yergenian, George, 195

Zinsser, Hans, 117